SpringerBriefs in Intelligent Systems

Artificial Intelligence, Multiagent Systems, and Cognitive Robotics

Series Editors

Gerhard Weiss, Maastricht University, Maastricht, The Netherlands

Karl Tuyls, University of Liverpool, Liverpool, UK; Google DeepMind London, UK

Editorial Board

Felix Brandt, Technische Universität München, Munich, Germany

Wolfram Burgard, Albert-Ludwigs-Universität Freiburg, Freiburg, Germany

Marco Dorigo, Université Libre de Bruxelles, Brussels, Belgium

Peter Flach, University of Bristol, Bristol, UK

Brian Gerkey, Open Source Robotics Foundation, Mountain View, USA

Nicholas R. Jennings, Imperial College London, London, UK

Michael Luck, King's College London, London, UK

Simon Parsons, City University of New York, New York, USA

Henri Prade, IRIT, Toulouse, France

Jeffrey S. Rosenschein, Hebrew University of Jerusalem, Jerusalem, Israel

Francesca Rossi, University of Padova, Padua, Italy

Carles Sierra, IIIA-CSIC Cerdanyola, Barcelona, Spain

Milind Tambe, University of Southern California, Los Angeles, USA

Makoto Yokoo, Kyushu University, Fukuoka, Japan

This series covers the entire research and application spectrum of intelligent systems, including artificial intelligence, multiagent systems, and cognitive robotics. Typical texts for publication in the series include, but are not limited to, state-of-the-art reviews, tutorials, summaries, introductions, surveys, and in-depth case and application studies of established or emerging fields and topics in the realm of computational intelligent systems. Essays exploring philosophical and societal issues raised by intelligent systems are also very welcome.

Chung-Chi Chen · Hiroya Takamura

Agent AI for Finance

From Financial Argument Mining to
Agent-Based Modeling

Chung-Chi Chen
Artificial Intelligence Research Center
AIST
Koto-ku, Tokyo, Japan

Hiroya Takamura
Artificial Intelligence Research Center
AIST
Koto-ku, Tokyo, Japan

ISSN 2196-548X ISSN 2196-5498 (electronic)
SpringerBriefs in Intelligent Systems
ISBN 978-3-031-94686-8 ISBN 978-3-031-94687-5 (eBook)
https://doi.org/10.1007/978-3-031-94687-5

This work was supported by National Institute of Advanced Industrial Science and Technology.

© The Editor(s) (if applicable) and The Author(s) 2025. This book is an open access publication.

Open Access This book is licensed under the terms of the Creative Commons Attribution 4.0 International License (http://creativecommons.org/licenses/by/4.0/), which permits use, sharing, adaptation, distribution and reproduction in any medium or format, as long as you give appropriate credit to the original author(s) and the source, provide a link to the Creative Commons license and indicate if changes were made.
The images or other third party material in this book are included in the book's Creative Commons license, unless indicated otherwise in a credit line to the material. If material is not included in the book's Creative Commons license and your intended use is not permitted by statutory regulation or exceeds the permitted use, you will need to obtain permission directly from the copyright holder.
The use of general descriptive names, registered names, trademarks, service marks, etc. in this publication does not imply, even in the absence of a specific statement, that such names are exempt from the relevant protective laws and regulations and therefore free for general use.
The publisher, the authors and the editors are safe to assume that the advice and information in this book are believed to be true and accurate at the date of publication. Neither the publisher nor the authors or the editors give a warranty, expressed or implied, with respect to the material contained herein or for any errors or omissions that may have been made. The publisher remains neutral with regard to jurisdictional claims in published maps and institutional affiliations.

This Springer imprint is published by the registered company Springer Nature Switzerland AG
The registered company address is: Gewerbestrasse 11, 6330 Cham, Switzerland

If disposing of this product, please recycle the paper.

This book is dedicated to all contributors in this field.

Preface

Financial documents contain numerous causal inferences and subjective opinions. In our previous book, "From Opinion Mining to Financial Argument Mining,"[1] we discussed understanding financial documents in a fine-grained manner, particularly those containing opinions. We highlighted several future directions, such as financial argument mining, multimodal opinion understanding, and analysis generation. We anticipated a lengthy journey for these topics. However, since 2022, ChatGPT and large language models (LLMs) have shown promising advancements, motivating us to write the second book that falls under the financial NLP topic. This book provides an overview of the current state of financial argument mining and financial text generation, and presents our thoughts on the blueprint for NLP in finance in the Agent AI era.

Agent-based AI systems have been widely discussed since the advent of LLMs. This book aims to equip researchers and practitioners with the latest methodologies, concepts, and frameworks for developing, deploying, and evaluating AI agents with capabilities in multimodal understanding, decision-making, and interaction. It places a special emphasis on human-centered decision-making and multi-agent cooperation in financial applications. We survey the current landscape and discuss future research and development directions.

Targeting a wide audience, from students to seasoned researchers in AI and finance, this book offers an overview of recent trends in Agent AI for finance. It provides a foundation for students to understand the field and design their research direction while inviting experienced researchers to engage in discussions on open research questions informed by pilot experimental results.

---

[1] Open Access: https://link.springer.com/book/10.1007/978-981-16-2881-8.

Although this book focuses on financial applications, the discussed concepts and methods can also be applied to other real-world applications by integrating domain-specific characteristics. We look forward to seeing new findings and more novel extensions based on the proposed ideas.[2]

Koto-ku, Japan
March 2025

Chung-Chi Chen
Hiroya Takamura

---

[2] We presented a tutorial at ECAI-2024 based on the content of this book. The slides are available at: https://sites.google.com/view/finagent/home.

Acknowledgments Many researchers have assisted us in writing this book. We would like to thank all the students and collaborators of the NLPFin—Prof. Hsin-Hsi Chen (National Taiwan University), Prof. Chenghua Lin (University of Manchester), Prof. Kiyoshi Izumi (University of Tokyo), Dr. Hen-Hsen Huang (Academia Sinica), Prof. Noriko Kando (National Institute of Informatics), Prof. Sudip Kumar Naskar (Jadavpur University), Prof. Yusuke Miyao (University of Tokyo), Prof. Ichiro Kobayashi (Ochanomizu University), Prof. Ryutaro Ichise (Tokyo Institute of Technology), Dr. Natthawut Kertkeidkachorn (Japan Advanced Institute of Science and Technology), Dr. Rungsiman Nararatwong (AIST), Dr. Ramon Ruiz Dolz (Universitat Politècnica de València), Jui Chu (National Taiwan University), Pei-Wei Kao (National Taiwan University), Tsun-Hsien Tang (National Taiwan University), JianTao Huang (National Taiwan University), Ting-Wei Hsu (National Taiwan University), Yi-Ting Liu (National Taiwan University), Ming-Xuan Shi (National Taiwan University), Chr-Jr Chiu (National Taiwan University), Wei-Lin Chen (National Taiwan University—AIST Intern), Sin-Han Yang (National Taiwan University), Tomas Goldsack (University of Sheffield—AIST Intern), Xingwei Qu (University of Manchester), Yuyang Cheng (University of Manchester), Yi-Ning Juan (National Taiwan University), Yu-Min Tseng (National Taiwan University), Chin-Yi Lin (National Taiwan University), Tsung-Hsuan Pan (National Taiwan University), Takehiro Takayanagi (The University of Tokyo—AIST RA), Hsiu-Hung Lee (National Yang Ming Chiao Tung University), Bo-Wei Chen (National Yang Ming Chiao Tung University), Hanwool Lee (NCSOFT), Yung-Yu Shih (National Taiwan University), Sohom Ghosh (Jadavpur University), Hsiu-Hsuan Yeh (National Taiwan University), Cheng-KuangWu (National Taiwan University), Prof. Yohei Seki (University of Tsukuba), Dr. Juyeon Kang (3DS Outscale), Anaïs Lhuissier (3DS Outscale), Prof. Min-Yuh Day (National Taipei University), and Prof. Yu-Lieh Huang (National Tsing Hua University).

On the funding side, this book was partially supported by a project JPNP20006, commissioned by the New Energy and Industrial Technology Development Organization (NEDO) and JSPS KAKENHI Grant Number 23K16956.

March 2025

Chung-Chi Chen
Hiroya Takamura

Competing Interests The authors have no competing interests to declare that are relevant to the content of this manuscript.

Contents

1 **Introduction** 1
   1.1   From Opinion Mining to Financial Argument Mining 1
   1.2   From Financial Argument Mining to Agent-Based Modeling 2
   1.3   Why Study Agent AI? 3
   1.4   Overview of the Book 7
   References 7

2 **Financial Argument Mining** 9
   2.1   Argument Structure 9
   2.2   Forward-Looking Argument Mining 12
   2.3   Argument Quality and Forecasting Skill Assessment 15
   2.4   Summary 17
   References 18

3 **Single-Agent/Model Design** 21
   3.1   Learning from Human Insights 21
   3.2   Retrieval-Augmented Generation 24
   3.3   Model Editing 27
   3.4   Summary 29
   References 30

4 **Multi-agent Interaction** 35
   4.1   Multi-round Discussion 35
   4.2   Hierarchical Decision-Making 38
   4.3   Human Behavior Simulacra 40
   4.4   Summary 43
   References 44

5	**Multi-scale Model Synergy**	47
	5.1 Data Augmentation	47
	5.2 Dynamic Interaction Loop	50
	5.3 Summary	53
	References	56
6	**Generative AI Application Scenarios**	59
	6.1 Extension of Impact Duration Inference	59
	6.2 Opinion Ranking	62
	6.3 Numeracy and Reasoning	64
	6.4 Creative Agent	66
	6.5 Summary	67
	References	68
7	**Looking to the Future**	71
	7.1 Progress on Previously Proposed Research Directions	71
	7.2 Future Directions	75
	7.3 Conclusion	80
	References	82

Chapter 1
Introduction

Intelligence encompasses understanding, reasoning, planning, inference, decision-making, and more. In our previous book [5], we focused on understanding financial documents, particularly those containing opinions, from an information extraction perspective. In this book, we extend our discussions to include argument-mining notions and further explore reasoning, planning, inference, and decision-making. This chapter provides an overview of the book. In Sect. 1.1, we recap the discussions from our previous book [5] and highlight the argument mining concept that will be discussed in this book. Section 1.2 outlines the roadmap from understanding human-written content to simulating human discussions. In Sect. 1.3, we share our thoughts on why Agent AI is a promising research topic for financial applications. Finally, we provide an overall structure of this book in Sect. 1.4.

1.1 From Opinion Mining to Financial Argument Mining

In our previous book [5], we proposed understanding financial opinions through the following twelve components: target entity, market sentiment, opinion holder, publishing time, validity period of an opinion, market information set, analysis aspect, degree of market sentiment, a set of claims, a set of premises, opinion quality, and influence power. Although we provided some results and promising directions at that time, several components remained unexplored, particularly the validity period, argument-based analysis, and opinion quality. In this book, we aim to fill this gap and share some experimental results. We will primarily use argument-mining notions for these discussions; thus, we first recap the idea of financial argument mining.

In contrast to opinion mining and sentiment analysis, which mainly focus on classifying a given statement as positive/negative (bullish/bearish), argument mining provides a more in-depth understanding of an opinion. For example, the causal relationship and the logic between premises and claims can provide clues for estimating the strength of an argument. Based on Toulmin's argumentative model [20], we can separate a narrative into two basic units, *claim* and *premise*, where the claim is the subjective view of the investor, and the premise is the objective act used to support the claim. With this idea, we can transform a financial analysis, such as a professional research report for a particular company, into directed graphs called argumentation structures. The argumentation structure has been widely used in other domains, such as debate quality assessment [15] and persuasive essay evaluation [21], but has yet to be discussed in financial analysis. In Chap. 2, we provide in-depth discussions on how to use argument-based analysis for financial purposes.

A major difference between financial arguments and arguments in other domains is that people always discuss the future in financial analysis. That is, we encounter many forward-looking statements in financial narratives. For example, investors discuss the company's future operations and possible stock price movements daily. This leads to two topics: the duration for which a given premise will influence companies' operations and the validity period of a given claim. For example, an event in 2020 may have little influence in 2025, and investors may not consider an opinion from 2019 when making decisions in 2026. Following this line of thought, we also propose adding a temporal dimension to the argument-based financial analysis in Chap. 2.

We believe that decomposing an analysis from the argument-mining perspective can lead to more insights for both human beings and machine learning models, and we hope our discussion will inspire more researchers to explore the potential of financial argument mining.

1.2 From Financial Argument Mining to Agent-Based Modeling

Simulating human behaviors and actions is one of the major goals of AI systems. After achieving the capability of understanding financial documents, generating (reasoning, planning, and inference), and making decisions becomes the next challenge. In 2021, it was still difficult for models to generate fluent financial analysis reports. Thanks to the development of generative models, we soon obtained superior LLMs for solving the fluency problem in 2022. In the past two years, researchers have started to discuss the potential of LLM agents, and our goal of generating analysis reports in our previous book [5] has become possible. Therefore, in Chaps. 3 and 4, we focus on how to use an agent or agents to simulate the behaviors of professionals in the financial sector, especially those writing analysis reports and making trading decisions.

The definition of agent-based AI systems (Agent AI) is still open to discussion. It has been defined as *"a class of interactive systems that can perceive visual stimuli, language inputs, and other environmentally grounded data, and can produce meaningful embodied actions"* [10]. In this book, we define Agent AI for finance as a system involving the interaction among multiple agents/models that can accept multimodal inputs (video, audio, image, and text) and is able to output useful information for the human-centered decision-making process in financial contexts. For example, when generating an analysis report based on the audio and transcript of an earnings conference call, several LLMs would be asked to examine different aspects, and then another LLM would be asked to summarize and make inferences based on the feedback from different LLMs. We discuss this kind of Agent AI in Chap. 4.

The definition of agent-based AI systems (Agent AI) is still open to discussion. It was defined as *"a class of interactive systems that can perceive visual stimuli, language inputs, and other environmentally grounded data, and can produce meaningful embodied actions"* [10]. In this book, we define Agent AI for finance as a system involving the interaction among multiple agents/models that can accept multimodal inputs (video, audio, image, and text) and is able to output useful information for the human-centered decision-making process. For example, when generating an analysis report based on the audio and transcript of an earnings conference call, several LLMs would be asked to examine different aspects, and then another LLM would be asked to summarize and make inferences based on the feedback from different LLMs. This is a kind of agent AI we will discuss in Chap. 4.

Agent-based modeling, which aims to simulate interactions in the real world, is a long-term topic of discussion in economics and finance [1]. Traditionally, agent-based modeling has always been done using several equations with different hyperparameter settings. With LLMs, we are exploring whether it is possible to conduct simulations in a more natural way. That is, we plan to let LLMs play different roles and interact with other LLMs using natural language. For example, we can simulate people's reactions to a rate hike in real estate and the stock market. If an LLM can simulate the decision of a given role, the interaction among LLMs would be very close to real-world outcomes. The same ideas can be applied to several tasks that were previously discussed with numerical simulation, such as consumer behavior and election simulation. Using LLMs for agent-based modeling enables researchers to observe changes in the simulated society and also provides explainable tracing routes for the changes. Although this is still in an early stage, we provide some discussions in Chap. 4.

1.3 Why Study Agent AI?

After explaining our journey and vision from opinion mining to agent-based modeling in finance, we now share our perspective on Agent AI and its promising future direction. Figure 1.1 illustrates our thoughts on this topic. AI models have traditionally been employed as tools for automation. For example, we can train a supervised model for sentiment analysis. By analyzing numerous social media posts, the model can determine the sentiments of social media users, effectively summarizing social media opinions [2]. Additionally, a model can be trained with annotated data to extract key information from financial documents, enabling investors to quickly grasp important

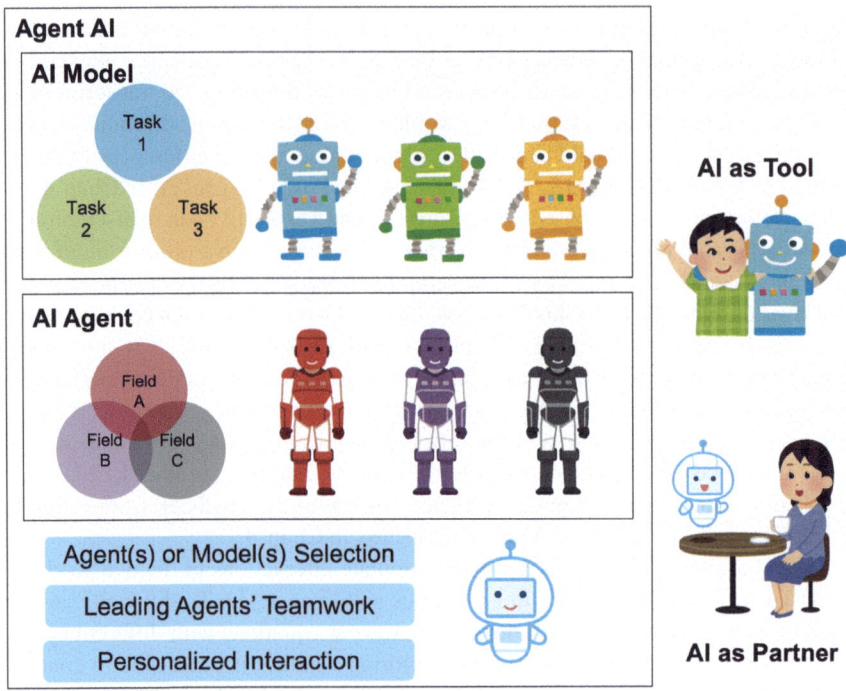

Fig. 1.1 Our vision of Agent AI

information [3, 4, 6, 7]. While these AI models may achieve high performance for specific tasks, they are generally limited to only one or a few tasks.

Unlike traditional AI models, LLMs acquire general capabilities after training on large datasets. In other words, AI models are analogous to a student who has specifically learned one chapter or skill, such as addition or a particular task. These models excel at solving problems within their specific domain because they have been trained solely for that purpose. For example, a model might be trained exclusively to perform addition, thereby performing exceptionally well when dealing with addition problems. In contrast, LLMs resemble a student who has received a broad foundational education, encompassing not only addition but also other mathematical concepts, and perhaps even knowledge from other subjects. These models possess a wider range of abilities, enabling them to handle various types of problems, although they may not always achieve perfect accuracy in every case. They are capable of answering open-ended questions in a variety of contexts, similar to how a student might apply their knowledge in different situations. Following this line of thought, it is conceivable that LLMs can be trained for different fields, much like students choosing a major in college.

Taking this a step further, intelligence encompasses not only natural language processing (NLP) but also the ability to plan, learn, make decisions, reason, and

1.3 Why Study Agent AI?

more. This leads to an extension of LLMs: the AI Agent. The concept of an AI agent is to empower LLMs to divide a task into several subtasks, iteratively check the generated text from different perspectives, select appropriate tools for solving problems, and so on. Unlike the one-step problem-solving approach of traditional AI models or LLMs, AI agents should simulate human behaviors in problem-solving, such as multi-step planning and reasoning. To implement a multi-step problem-solving framework, an admin agent is required to select AI agents to form a team and guide the discussions among them. The admin agent also communicates with users and may require personalization capabilities. This multi-step framework signifies a transition from the "AI as Tool" era to the "AI as Partner" era, as illustrated in Fig. 1.1. Specifically, when AI is seen as a tool, it is primarily used to accomplish specific tasks. In this mode, AI executes tasks assigned by humans but doesn't operate beyond the given instructions. Humans rely on AI for automation, but ultimate decision-making and creativity remain in human control. For example, spell check or grammar correction in writing software can detect errors, but humans decide whether to accept the AI's suggestions. AI in photo editing tools can automatically enhance photos based on patterns, but the user has the final say on adjustments and edits. However, when AI is seen as a partner, it plays a more proactive and collaborative role. AI not only executes tasks but also participates in the thinking and decision-making process. It helps solve creative and strategic problems, acting more as a co-worker than just a tool. For example, in healthcare, an AI partner provides diagnostic suggestions and treatment options based on patient history and the latest medical research, and doctors and AI collaborate to find the best treatment plan. In financial analysis, an AI partner doesn't just analyze data but also offers insights into market trends, helps executives predict future outcomes, and guides decision-making collaboratively. In sum, "AI as a tool" is limited to executing tasks assigned by humans, focusing on specific operations without influencing decisions or creative processes. "AI as a partner" works alongside humans, engaging in decision-making, offering creative input, and contributing to strategic problem-solving in a collaborative manner.

Recent studies have demonstrated the potential of the multi-agent framework. For example, instead of relying on a single LLM to generate code, employing multiple agents to work collaboratively can yield better outcomes in software development [12]. These agents assume roles such as product managers, architects, project managers, and engineers and interact with one another to refine and generate code based on the given software requirements. Another significant application scenario is human behavior simulation [16]. This includes simulating the daily life of a small village or human behaviors in various professional contexts, such as buyer-seller interactions. By accurately capturing the persona of each market participant, we can approach a realistic market simulation. Furthermore, it is possible to observe whether AI agents exhibit behaviors similar to humans, such as overconfidence [19], overreaction [9], or herding behavior [8]. AI agents introduce a new dimension to agent-based modeling, as they can now communicate using natural language rather than the numerical methods previously employed. This advancement opens up opportunities to revisit a wide range of agent-based modeling research by employing AI agents.

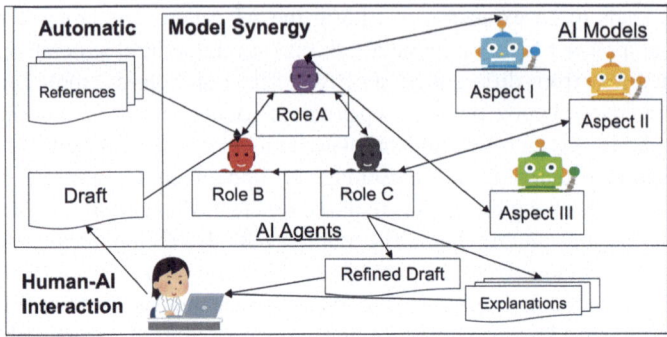

Fig. 1.2 Illustration of multi-scale model synergy

In addition to discussions focused solely on LLM-based agent interaction [11], we propose the concept of multi-scale model synergy in Chap. 5. Our goal is to bridge the efforts on pre-trained language models (PLMs) and LLMs to develop an advanced multi-agent framework. As illustrated in Fig. 1.1, previous studies have primarily concentrated on either AI model construction or AI agent exploration, with multi-agent frameworks largely relying on LLMs. However, supervised models often outperform LLMs on specific tasks. It is imprudent to disregard these models in the LLM era. Therefore, we propose that admin agents should not only select agents but also choose models for various checks or evaluations of the agents' outputs from multiple perspectives. Additionally, agents can utilize feedback from these models to refine their outputs further.

Figure 1.2 presents an example. Imagine that we are now company managers preparing a speech script for an earnings conference call. We can utilize an LLM to generate the script by providing a set of references. Additionally, we can rely on AI agents to discuss various aspects, such as compliance and professionalism. However, it is also important to consider the market reaction to the speech script. Several models are specifically designed to predict market reactions based on earnings calls [14, 17, 18]. Given that these models perform better than agents, it is necessary to integrate them into the multi-agent framework. This is why we propose the concept of multi-scale model synergy. Furthermore, managers may need to anticipate the types of questions that professional analysts might ask [13], and models fine-tuned for this purpose should be adopted rather than relying solely on prompting LLMs. More discussions will be provided in Chap. 5.

These are the reasons we believe that agent AI is a promising direction. In this book, we will share findings based on previous explorations and further discuss how to advance the concept of multi-scale model synergy.

1.4 Overview of the Book

Chapter 2 connects our previous work with this book by providing a clearer understanding of financial argument mining. This chapter also presents some of our explorations on the topic, and the results reveal the importance of incorporating argument-mining concepts into financial NLP. In Chap. 3, we begin the discussion on single-agent design and offer a survey of recent studies, covering topics such as multimodal agents, feature engineering and promotion, model editing, human annotation and feedback, and retrieval-augmented generation (RAG). Chapter 4 outlines the current state of multi-agent interaction development, sharing insights and findings related to behavior simulation, trading decision-making, and multimodal multi-agent interaction. In Chap. 5, we discuss the proposed multi-scale model synergy and identify several open research questions. Chapter 6 explores additional use cases of argument mining in finance and agent AI designed for finance. Finally, we highlight future research directions for agent AI in finance and conclude the book.

References

1. AXTELL, R. L., AND FARMER, J. D. Agent-based modeling in economics and finance: Past, present, and future. *Journal of Economic Literature* (2022), 1–101.
2. BOLLEN, J., MAO, H., AND ZENG, X. Twitter mood predicts the stock market. *Journal of Computational Science 2*, 1 (2011), 1–8.
3. CHEN, C.-C., HUANG, H.-H., AND CHEN, H.-H. Crowd View: Converting investors' opinions into indicators. In *IJCAI* (2019), pp. 6500–6502.
4. CHEN, C.-C., HUANG, H.-H., AND CHEN, H.-H. Numeral attachment with auxiliary tasks. In *Proceedings of the Forty-Second International ACM SIGIR Conference on Research and Development in Information Retrieval* (2019), pp. 1161–1164.
5. CHEN, C.-C., HUANG, H.-H., AND CHEN, H.-H. *From opinion mining to financial argument mining*. Springer Nature, 2021.
6. CHEN, C.-C., HUANG, H.-H., SHIUE, Y.-T., AND CHEN, H.-H. Numeral understanding in financial tweets for fine-grained crowd-based forecasting. In *2018 IEEE/WIC/ACM International Conference on Web Intelligence (WI)* (2018), IEEE, pp. 136–143.
7. CHEN, C.-C., HUANG, H.-H., TSAI, C.-W., AND CHEN, H.-H. CrowdPT: Summarizing crowd opinions as professional analyst. In *The World Wide Web Conference* (2019), pp. 3498–3502.
8. CLEMENT, M. B., AND TSE, S. Y. Financial analyst characteristics and herding behavior in forecasting. *The Journal of finance 60*, 1 (2005), 307–341.
9. DE BONDT, W. F., AND THALER, R. Does the stock market overreact? *The Journal of finance 40*, 3 (1985), 793–805.
10. DURANTE, Z., HUANG, Q., WAKE, N., GONG, R., PARK, J. S., SARKAR, B., TAORI, R., NODA, Y., TERZOPOULOS, D., CHOI, Y., ET AL. Agent AI: Surveying the horizons of multimodal interaction. *arXiv preprint* arXiv:2401.03568 (2024).
11. GUO, T., CHEN, X., WANG, Y., CHANG, R., PEI, S., CHAWLA, N., WIEST, O., AND ZHANG, X. Large language model based multi-agents: A survey of progress and challenges. In *33rd International Joint Conference on Artificial Intelligence (IJCAI 2024)* (2024), IJCAI.
12. HONG, S., ZHUGE, M., CHEN, J., ZHENG, X., CHENG, Y., WANG, J., ZHANG, C., WANG, Z., YAU, S. K. S., LIN, Z., ET AL. Metagpt: Meta programming for a multi-agent collaborative framework. In *The Twelfth International Conference on Learning Representations*.

13. Juan, Y., Chen, C.-C., Huang, H.-H., and Chen, H.-H. Generating multiple questions from presentation transcripts: A pilot study on earnings conference calls. In *Proceedings of the 16th International Natural Language Generation Conference* (Prague, Czechia, Sept. 2023), C. M. Keet, H.-Y. Lee, and S. Zarrieß, Eds., Association for Computational Linguistics, pp. 449–454.
14. Koval, R., Andrews, N., and Yan, X. Forecasting earnings surprises from conference call transcripts. In *Findings of the Association for Computational Linguistics: ACL 2023* (Toronto, Canada, July 2023), A. Rogers, J. Boyd-Graber, and N. Okazaki, Eds., Association for Computational Linguistics, pp. 8197–8209.
15. Li, J., Durmus, E., and Cardie, C. Exploring the role of argument structure in online debate persuasion. In *Proceedings of the 2020 Conference on Empirical Methods in Natural Language Processing (EMNLP)* (Online, Nov. 2020), Association for Computational Linguistics, pp. 8905–8912.
16. Park, J. S., O'Brien, J., Cai, C. J., Morris, M. R., Liang, P., and Bernstein, M. S. Generative agents: Interactive simulacra of human behavior. In *Proceedings of the 36th Annual ACM Symposium on User Interface Software and Technology* (New York, NY, USA, 2023), UIST '23, Association for Computing Machinery.
17. Qin, Y., and Yang, Y. What you say and how you say it matters: Predicting stock volatility using verbal and vocal cues. In *Proceedings of the Fifty-Seventh Annual Meeting of the Association for Computational Linguistics* (Florence, Italy, July 2019), Association for Computational Linguistics, pp. 390–401.
18. Sang, Y., and Bao, Y. DialogueGAT: A graph attention network for financial risk prediction by modeling the dialogues in earnings conference calls. In *Findings of the Association for Computational Linguistics: EMNLP 2022* (Abu Dhabi, United Arab Emirates, Dec. 2022), Y. Goldberg, Z. Kozareva, and Y. Zhang, Eds., Association for Computational Linguistics, pp. 1623–1633.
19. Skala, D. Overconfidence in psychology and finance-an interdisciplinary literature review. *Bank I kredyt*, 4 (2008), 33–50.
20. Toulmin, S. E. *The Uses of Argument*. Cambridge University Press, 2003.
21. Wachsmuth, H., Al Khatib, K., and Stein, B. Using argument mining to assess the argumentation quality of essays. In *Proceedings of COLING 2016, the Twenty-Sixth International Conference on Computational Linguistics: Technical Papers* (2016), pp. 1680–1691.

Open Access This chapter is licensed under the terms of the Creative Commons Attribution 4.0 International License (http://creativecommons.org/licenses/by/4.0/), which permits use, sharing, adaptation, distribution and reproduction in any medium or format, as long as you give appropriate credit to the original author(s) and the source, provide a link to the Creative Commons license and indicate if changes were made.

The images or other third party material in this chapter are included in the chapter's Creative Commons license, unless indicated otherwise in a credit line to the material. If material is not included in the chapter's Creative Commons license and your intended use is not permitted by statutory regulation or exceeds the permitted use, you will need to obtain permission directly from the copyright holder.

Chapter 2
Financial Argument Mining

This chapter focuses on financial argument mining. In Sect. 2.1, we first recap the concept of the argument structure of financial opinions, which we discussed in our previous work [6]. We also discuss the statistics of argument structures of company managers and professional analysts based on the annotated data. In Sect. 2.2, we introduce an extended concept, forward-looking argument mining, by presenting the ideas of "scenario" and "impact duration." The demonstration of how to apply the concept of forward-looking argument mining for downstream tasks is provided in Sect. 2.3. Finally, we conclude this chapter with a summary in Sect. 2.4.

2.1 Argument Structure

In Sects. 2.2 and 2.3 of our previous book [6], we introduced the concept of converting raw opinions from investors into an argument structure. This encompasses both the internal structure of a single opinion and the relationships among multiple opinions. Rather than simply presenting the raw analysis from a professional analyst, we propose breaking it down into distinct argument units and further understanding it in a structured form. These argument units can be categorized into claims and premises, and the final recommendation (e.g., Overweight) is referred to as the main claim. Additionally, the connections between argument units vary in weight and quality. Some reasons may strongly support the claim, while others may be weaker due to the lack of a causal relationship. The overall quality of a claim is also influenced by the premises and inferences supporting it.

These concepts have been previously explored and discussed. Recently, we have constructed datasets based on these concepts, including stock analysis reports [17] and earnings conference calls [3]. Although we identified four primary sources for financial NLP [6]—managers, professionals, social media users, and journals— extracting premises and claims from the latter two is challenging. Social media users, particularly those on Twitter/X, rarely provide premises to support their claims [4], making social media data less suitable for this research direction. Similarly, journalists typically report events without expressing their opinions, as seen in news articles. This renders such articles an inappropriate target for this research. Table 2.1 presents the statistics of these two existing datasets. In addition to argument units, we further propose the use of different sentiment labels for premises and claims in professional analysis. Positive/negative sentiment, as used in traditional sentiment analysis, should describe whether the events that occurred are good or bad, which is appropriate for premises. However, claims represent subjective opinions, and in financial contexts, they should be labeled as bullish or bearish to reflect the investor's perspective. Therefore, we propose using bullish/bearish to indicate the market sentiment of claims. The statistics highlight the distinct narrative styles of analysts and managers. Analysts tend to use multiple pieces of evidence to support a single claim, while managers generally provide just one premise per claim.

In argumentative narratives, the author or speaker aims to persuade the readers or listeners to accept their viewpoint. This is also true in managers' speeches, where they not only disclose news and operational updates but also attempt to persuade investors to hold or buy their company's stock. Although argumentation strategies in various documents, such as editorials [1], Wikipedia articles [2], and online debate discussions [19], have been analyzed, few studies offer fine-grained annotations for examining the argumentation strategies of managers and investors. By adding additional annotations to the existing dataset [8], we provide an overview of managers' speeches in Table 2.2. To facilitate comparison, the argumentation strategies of professional analysts' reports are shown in Table 2.3.

Table 2.1 Statistics of financial argument mining datasets [9]

	Professional research reports			Earnings conference calls
	Positive/bullish (%)	Neutral (%)	Negative/bearish (%)	–
Premise	29.68	6.62	26.30	52.40
Claim	16.29	6.59	14.51	47.60
Total	17,214			9691

Table 2.2 Statistics of managers' argumentation strategy

		Premise		
		Positive (%)	Neutral (%)	Negative (%)
Claim	Positive	2.58	5.17	0.00
	Neutral	2.54	87.26	0.94
	Negative	0.00	0.40	1.11

2.1 Argument Structure

Table 2.3 Statistics of investors' argumentation strategy

		Premise		
		Positive (%)	Neutral (%)	Negative (%)
Claim	Bullish	24.17	1.79	0.00
	Neutral	0.19	67.67	0.04
	Bearish	1.17	0.31	4.66

On the basis of these annotations, we have derived several findings. First, managers' speeches contain many neutral narratives, with only a few descriptions exhibiting sentiment polarity. In contrast, approximately 30% of investors' narratives exhibit sentiment polarity, with positive arguments being more prevalent than negative ones. Second, managers never use an optimistic premise to support a negative claim, nor do they use a pessimistic premise to support a positive claim. However, in investors' narratives, some bearish claims are associated with positive premises, which indicates that positive news does not always lead investors to make bullish claims. Third, when presented with a positive or negative premise, managers tend to conclude in a neutral tone, whereas investors consistently make either a bullish or bearish claim. Although both managers and professional analysts aim to persuade investors to make investment decisions based on their viewpoints, these findings reveal significant differences in their narratives.

We further provide an analysis of argumentation structures in both managers' speeches and analysts' reports. It is important to note that the argumentation structure discussed in this section is paragraph-based, meaning it concerns the order and relationships of claims and premises within the same paragraph.

Given that a claim may be supported by several premises, and a premise can support more than one claim, we observe that 14.09% of claims are supported by multiple premises, while 46.75% of premises support more than one claim. Following this observation, we identify three commonly used structures: (1) premises are provided before claims (P → C), (2) a claim is made before providing premises to support it (C ← P), and (3) premises are presented, followed by a claim with additional premises (P → C ← P). Tables 2.4 and 2.5 present the statistics of these structures in both narratives.

Table 2.4 Statistics of managers' argumentation structure

		Structure		
		$P \to C$ (%)	$C \leftarrow P$ (%)	$P \to C \leftarrow P$ (%)
Claim	Positive	94.29	5.71	0.00
	Neutral	96.43	3.57	0.00
	Negative	87.06	12.65	0.23

Table 2.5 Statistics of investors' argumentation structure

		Structure		
		P → C (%)	C ← P (%)	P → C ← P (%)
Claim	Bullish	72.60	21.46	5.94
	Neutral	62.21	29.13	8.66
	Bearish	100.00	0.00	0.00

Our findings indicate that both managers and investors tend to use the structure where premises are provided before claims, while the (P → C ← P) structure is rarely employed. Additionally, we find that professional investors use the (C ← P) structure more frequently than managers. This suggests that some analysts prefer to state their opinions before providing explanations. Notably, when making negative claims, analysts consistently use the (P → C) structure in their narratives. Our analysis offers a novel perspective for investigating financial narratives and suggests that fine-grained argument understanding could enhance downstream applications, such as stock movement prediction.

In summary, we demonstrate the difference between company managers' talks and professional analysts' reports from an argument-mining perspective. Based on the results of several studies [9, 17], identifying argument units in financial narratives is not particularly challenging. The follow-up question is: how can these argument features be utilized for downstream tasks? We provide some examples in the following sections.

2.2 Forward-Looking Argument Mining

As mentioned in the previous section, argument mining, which provides a more fine-grained analysis of opinions, is one of the emerging topics in opinion mining and sentiment analysis research. Recent studies have primarily focused on summarizing past and current evidence and forming persuasive expressions by selecting appropriate structures and strategies to support a given stance. We contend that arguments with a forward-looking and future-oriented perspective have received little attention in previous research. However, forward-looking statements are both commonly used and important in daily life. For example, discussions on who will win the next presidential election and which stock will outperform others are typical real-life scenarios. Instead of making predictions, our goal is to provide rational explanations within the framework of argument mining. To this end, this section proposes a research agenda for forward-looking argument mining. It encompasses research issues from understanding to generating forward-looking arguments.

Figure 2.1 illustrates an example of the proposed idea, specifically an instance from the analysis report. To comprehend the opinion, we first need to separate it

2.2 Forward-Looking Argument Mining

In response to the impact of the China-US trade war, the bicycle companies, Merida and Giant, transferred some US orders from China to Taiwan, but faced severe labor shortages and insufficient production capacity. Taiwan is not only difficult to recruit, but also has relatively high personnel costs. At the same time, considering that 1) the trade war **may** escalate to 25% uncertainty; 2) the negative impact of China's bicycle sharing **will lead** to weak sales performance; 3) the low utilization rate of China's factories **will make** gross Interest rates continue to be under pressure, so **we keep neutral attitude** on Giant and Merida in 2019.

- Premise
- Scenario
- Forward-Looking Claim

Fig. 2.1 Example of forward-looking argument mining

into three components: premise, scenario, and forward-looking claim. In addition to extending traditional argument identification tasks, we propose two implicit information assessment tasks: estimating the impact period of the opinion and evaluating the supporting strength of the premises. In summary, the novelty of this concept is threefold: (1) integrating scenario planning concepts into argument mining, (2) assessing the impact period of arguments, and (3) evaluating the supporting strength of forward-looking arguments.

In financial analysis, individuals frequently discuss future possibilities based on available information. This indicates that relying solely on current premises may be insufficient to fully comprehend analyses from managers or analysts. They often combine projections of future events with past occurrences. Scenario planning is also prevalent in everyday contexts. For example, in debates about "climate change," one might argue that "rising sea levels and coastal flooding could displace many people"—a description of possible future actions or events. We argue that the modern era, characterized by rapid technological advancements, geopolitical instability, and evolving socio-economic landscapes, embodies the VUCA paradigm: Volatile, Uncertain, Complex, and Ambiguous. In such a context, traditional predictive methods, which often rely on linear extrapolations of current trends, are insufficient.

Unlike predictive models, which aim for precision by analyzing historical and present data, scenario planning explores a different domain. It does not merely extrapolate a single possible future but instead constructs multiple narratives, each illustrating a potential outcome. Scenario planning acknowledges the multifaceted and uncertain nature of the future, accounting for various driving forces and uncertainties to develop alternative future scenarios, some of which may seem unlikely. The objective is not to predict the "correct" future but to prepare for a range of possibilities.

Scenario planning is particularly useful in fields marked by uncertainty. Long-term strategic planning, geopolitics, and emerging industries are areas where traditional linear forecasting falls short. Although scenario planning is not a new concept [12], its integration with modern technological tools presents exciting opportunities. Recent research [13, 20] highlights the growing potential for refining, automating, and diversifying scenario planning through NLP. The ability of NLP to analyze extensive textual datasets, identify emerging patterns, and generate detailed narratives makes it

Table 2.6 Statistics of scenario and impact duration annotations [17]

Scenario			Impact duration		
Continued growth	3330	46.82%	<1 month	769	2.43%
Steady state	670	9.42%	1–3 month	6886	21.75%
Collapse	2558	35.96%	4–6 month	4445	14.04%
Transformation	555	7.80%	7–12 month	12,125	38.30%
			>1 year	7431	23.47%

a powerful asset for scenario planners [10, 21]. In line with this, we contend that, especially for financial analysis, understanding and generating scenarios represents the next crucial step in advancing NLP methodologies.

To go one step further, when discussing past events, we already know when they occurred, but when proposing potential scenarios, the impact duration can vary. For example, some may argue that the announcement of the Pixel 9 Pro will significantly influence Google's phone department revenue for half a year, while others might believe the impact will only last a quarter. This variation depends on how analysts interpret the situation and the scenarios they envision. How experts estimate impact duration and how models can simulate expert estimations remain open questions on this topic.

To address this, Lin et al. [17] introduced a dataset, Equity-AMSA, a dataset containing approximately 7000 and 24,000 annotations for scenarios and impact duration estimations in equality analysis reports. Table 2.6 presents the statistics of their labels. First, analysts' scenarios generally fall into two categories: continued growth and collapse. It is reasonable that they release reports for these important events instead of those that may have a minor impact on the company's operation. Second, it is notable that professionals tend to focus on company operations extending beyond one month, which differs from the typical stock movement prediction timeframe of 30 d frequently used in research [18, 22].

Previous studies have attempted to predict short-term price movements, such as one-day or three-day returns, which could be categorized as speculation—trading for quick profits with the highest risk and uncertainty. Conversely, few studies in the AI field have tackled the issue of investment. The aim of investment is to earn returns from the market while managing proper risk. As indicated in the empirical studies of the 2013 Nobel Memorial Prize in Economic Sciences laureates Eugene Fama, Lars Peter Hansen, and Robert Shiller, it is impossible to predict short-term price movements, whereas long-term forecasts are feasible.

We seek to differentiate speculation from investment. Unlike speculation, which focuses on analyzing information embedded in historical prices, investment focuses on fundamental factors such as growth rates and default rates. When targeting different financial instruments, investors consider varying information to evaluate the value of the instrument. For example, when analyzing stocks, investors may focus on the growth rate; for company bonds, the default rate is critical; and for foreign exchange, the interest rate may be a key factor. We believe that forward-looking argument mining could become an important topic as we shift from speculation to investment.

2.3 Argument Quality and Forecasting Skill Assessment

Given that forward-looking arguments are all related to the future, they often involve forecasts for future events. An important question is whether these forecasts will come true. We argue that this may be related to the premises and scenarios used by the opinion holder. Specifically, if a claim is supported by an incorrect causal inference, it could be considered coincidental if it proves accurate in the future. When presented with two claims (e.g., a bullish view) supported by different premises, estimating which one is stronger becomes a challenge. Similarly, determining which of two conflicting claims to follow is also difficult. This raises the need for a method to evaluate forecasting skill. To this end, we propose an approach based on the framework of forward-looking argument concepts.

In the previous section, we introduced the concept of separating an opinion into three components and discussed the task of impact duration estimation. One question remains from Fig. 2.1: the strength of the supporting evidence. Argument mining has proven useful for various downstream applications in different fields. For example, encoding argument structures into models improves the performance of predicting which debater presents a more convincing argument on online debate platforms [16]. It is also beneficial for automatic essay scoring [14]. Although financial documents have been extensively utilized in different experiments, few previous works discuss the role of arguments in financial narratives. Most current studies analyzing financial narratives rely on traditional features such as sentiment, readability, and parts of speech [15, 23]. However, evaluating the strength of the supporting premises remains largely unexplored and presents a particular challenge, especially in forward-looking analysis. In Fig. 2.2, we propose using the difference between the price target set by professional analysts and the closing price on the report release day as a proxy for

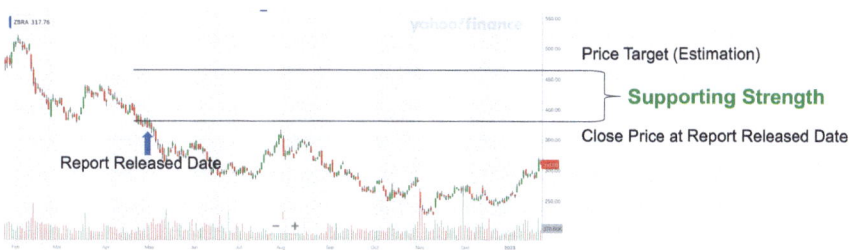

Fig. 2.2 Example of supporting strength estimation

Table 2.7 Analysts' argument strengths with different narratives

	Claim (%)	Premise (%)
Negative	18.31	17.98
Without negative	21.53	21.52

argument strength. The rationale is that analysts will provide a higher price target if they believe the stock is undervalued, and the supporting reasons in these reports are likely stronger than those in reports with lower price targets.

We offer an example to illustrate whether analysts' argumentation strategies reflect their supporting strengths. Table 2.7 presents the results of our statistical analysis. We find that the strengths of reports containing negative claims or premises is approximately 3% lower than that of reports containing only positive and neutral arguments, and that negative premises have a greater impact than negative claims. These findings suggest a potential direction for future research, namely, combining fine-grained sentiment analysis with argument mining in the analysis of forward-looking statements.

We further adopt the concept of argument strength for ranking 2280 professional analysts' reports [7]. More specifically, we train the BERT model with sentence-level argument strength estimation and then let the model assess the argument strength of all sentences in all reports. The average score of the sentences in a report is used to rank all reports. We adopt the maximum possible profit (MPP) and maximum loss (ML) metrics for evaluation, as in previous work [5]. To recap this concept, for bullish and bearish opinions posted on day t, we calculate the maximum possible profit (MPP) and maximum loss (ML) as follows:

$$MPP_{bullish} = \frac{\max_{i=t+1}^{t+T} H_i - O_{t+1}}{O_{t+1}}, \tag{2.1}$$

$$ML_{bullish} = \frac{\min_{i=t+1}^{t+T} L_i - O_{t+1}}{O_{t+1}}, \tag{2.2}$$

$$MPP_{bearish} = \frac{O_{t+1} - \min_{i=t+1}^{t+T} L_i}{O_{t+1}}, \tag{2.3}$$

$$ML_{bearish} = \frac{O_{t+1} - \max_{i=t+1}^{t+T} H_i}{O_{t+1}}, \tag{2.4}$$

where O_t denotes the opening price on day t, H_t denotes the highest prices on day t, L_t denotes the lowest prices on day t, and T is a given integer.

Table 2.8 shows the potential of learning support strength using the price target as a proxy. First, the MPP of the top 20% of reports is higher than that of the bottom 20%. Second, while the MPP is higher, the ML is also higher for the top-ranked reports. This is reasonable, as pursuing higher profits is correlated with higher risks. Overall,

Table 2.8 MPP and ML of the reports ranked based on argument strength

	Top		Last	
	10th decile (%)	9th decile (%)	2nd decile (%)	1st decile (%)
MPP	15.25	14.53	12.75	11.93
ML	−11.30	−11.98	−9.24	−8.80

Table 2.9 Experimental results of GNN with different features for forecasting skill assessment [17]

Argument units	Argument-based sentiment analysis	Impact duration	Acc.	F1-Score
–	–	–	0.731	0.728
✓	–	–	0.749	0.741
✓	✓	–	0.775	0.773
✓	✓	✓	0.798	0.796

these results indicate that it is possible to perform argument quality assessment via the proposed forward-looking argument mining notions. However, although we link quality with profitability, there remains room for improvement in quality assessment tasks in a more fine-grained manner.

Going one step further, we also find that the proposed argument mining notion is useful for forecasting skill assessment [17]. Specifically, we find that we can estimate whether the price target will be achieved by incorporating argument-based sentiment analysis results into the graph neural network (GNN) model [11]. Table 2.9 shows the experimental results, supporting the usefulness of the proposed financial argument mining concepts.

In summary, this section introduces two important but rarely discussed tasks in financial opinion mining, providing a pilot exploration from the perspective of fine-grained argument mining. Many combinations could be explored in the future, and we hope the discussion in this section offers insights for future studies on forward-looking argument analysis.

2.4 Summary

In this chapter, we explored the evolving landscape of financial argument mining, and emphasized both established and emerging methodologies. We began by recapping the core argument structure used in financial narratives, and demonstrated how it applies to professional analyst reports and company manager communications. These structures allow for a clearer, more systematic understanding of the relationships between claims, premises, and recommendations.

Through the data analysis presented, we clarified the significant differences between the argumentation strategies of company managers and professional analysts Managers tend to focus on providing neutral narratives with fewer sentiment-laden premises, while analysts often express stronger views, with more sentiment-driven arguments, particularly bullish or bearish claims. These differences highlight the distinct communicative goals of managers and analysts, where managers aim to stabilize perceptions of their company's performance, and analysts often engage in more direct recommendations to investors. We then introduced the concept of forward-looking argument mining, which goes beyond conventional argument analysis by incorporating scenario planning and impact duration estimation. This extension acknowledges the importance of considering not just what has happened but what might happen, especially in the uncertain context of financial markets. We also discussed how argument strength can be evaluated with particular attention to forward-looking financial opinions. Lastly, we proposed practical applications of these concepts for assessing forecasting skills and argument quality, particularly within the framework of predicting stock movements and evaluating the strength of professional financial reports. We demonstrated the potential benefits of integrating fine-grained argument mining with financial sentiment analysis. We also highlighted its relevance for both short-term trading and long-term investment decisions. Additional discussions on argument quality assessment are presented in Sect. 6.2.

In summary, this chapter presents a comprehensive framework for financial argument mining. The applications are extended into various downstream tasks such as stock movement prediction and quality assessment of financial analyses. The fine-grained understanding of financial narratives offered by argument mining holds great promise for improving financial decision-making and predictive capabilities in the ever-complex world of financial markets.

References

1. AL-KHATIB, K., WACHSMUTH, H., KIESEL, J., HAGEN, M., AND STEIN, B. A news editorial corpus for mining argumentation strategies. In *Proceedings of COLING 2016, the 26th International Conference on Computational Linguistics: Technical Papers* (Osaka, Japan, Dec. 2016), The COLING 2016 Organizing Committee, pp. 3433–3443.
2. AL-KHATIB, K., WACHSMUTH, H., LANG, K., HERPEL, J., HAGEN, M., AND STEIN, B. Modeling deliberative argumentation strategies on Wikipedia. In *Proceedings of the 56th Annual Meeting of the Association for Computational Linguistics (Volume 1: Long Papers)* (Melbourne, Australia, July 2018), Association for Computational Linguistics, pp. 2545–2555.
3. ALHAMZEH, A., FONCK, R., VERSMÉE, E., EGYED-ZSIGMOND, E., KOSCH, H., AND BRUNIE, L. It's time to reason: Annotating argumentation structures in financial earnings calls: The finarg dataset. In *Proceedings of the Fourth Workshop on Financial Technology and Natural Language Processing (FinNLP)* (2022), pp. 163–169.
4. CHEN, C.-C., HUANG, H.-H., AND CHEN, H.-H. Issues and perspectives from 10,000 annotated financial social media data. In *Proceedings of the Twelfth Language Resources and Evaluation Conference* (Marseille, France, May 2020), N. Calzolari, F. Béchet, P. Blache, K. Choukri, C. Cieri, T. Declerck, S. Goggi, H. Isahara, B. Maegaard, J. Mariani, H. Mazo, A. Moreno, J. Odijk, and S. Piperidis, Eds., European Language Resources Association, pp. 6106–6110.

References

5. CHEN, C.-C., HUANG, H.-H., AND CHEN, H.-H. Evaluating the rationales of amateur investors. In *The World Wide Web Conference* (2021).
6. CHEN, C.-C., HUANG, H.-H., AND CHEN, H.-H. *From opinion mining to financial argument mining*. Springer Nature, 2021.
7. CHEN, C.-C., HUANG, H.-H., CHEN, H.-H., TAKAMURA, H., KOBAYASHI, I., AND MIYAO, Y. Enhancing investment opinion ranking through argument-based sentiment analysis. In *arXiv* (2024).
8. CHEN, C.-C., HUANG, H.-H., HUANG, Y.-L., AND CHEN, H.-H. Distilling numeral information for volatility forecasting. In *Proceedings of the 30th ACM International Conference on Information & Knowledge Management* (2021), pp. 2920–2924.
9. CHEN, C.-C., LIN, C.-Y., CHIU, C.-J., HUANG, H.-H., ALHAMZEH, A., HUANG, Y.-L., TAKAMURA, H., AND CHEN, H.-H. Overview of the ntcir-17 finarg-1 task: Fine-grained argument understanding in financial analysis. In *Proceedings of the 17th NTCIR Conference on Evaluation of Information Access Technologies, Tokyo, Japan* (2023), pp. 12–15.
10. FEBLOWITZ, M., HASSANZADEH, O., KATZ, M., SOHRABI, S., SRINIVAS, K., AND UDREA, O. Ibm scenario planning advisor: a neuro-symbolic erm solution. In *Proceedings of the AAAI Conference on Artificial Intelligence* (2021), vol. 35, pp. 16032–16034.
11. HAMILTON, W., YING, Z., AND LESKOVEC, J. Inductive representation learning on large graphs. *Advances in neural information processing systems 30* (2017).
12. HASHIMOTO, C., TORISAWA, K., KLOETZER, J., SANO, M., VARGA, I., OH, J.-H., AND KIDAWARA, Y. Toward future scenario generation: Extracting event causality exploiting semantic relation, context, and association features. In *Proceedings of the 52nd Annual Meeting of the Association for Computational Linguistics (Volume 1: Long Papers)* (2014), pp. 987–997.
13. ISHIGAKI, T., NISHINO, S., WASHINO, S., IGARASHI, H., NAGAI, Y., WASHIDA, Y., AND MURAI, A. Automating horizon scanning in future studies. In *Proceedings of the Thirteenth Language Resources and Evaluation Conference* (2022), pp. 319–327.
14. JIANG, S., YANG, K., SUVARNA, C., CASULA, P., ZHANG, M., AND ROSÉ, C. Applying Rhetorical Structure Theory to student essays for providing automated writing feedback. In *Proceedings of the Workshop on Discourse Relation Parsing and Treebanking 2019* (Minneapolis, MN, June 2019), Association for Computational Linguistics, pp. 163–168.
15. KEITH, K., AND STENT, A. Modeling financial analysts' decision making via the pragmatics and semantics of earnings calls. In *Proceedings of the Fifty-Seventh Annual Meeting of the Association for Computational Linguistics* (2019), pp. 493–503.
16. LI, J., DURMUS, E., AND CARDIE, C. Exploring the role of argument structure in online debate persuasion. In *Proceedings of the 2020 Conference on Empirical Methods in Natural Language Processing (EMNLP)* (Online, Nov. 2020), Association for Computational Linguistics, pp. 8905–8912.
17. LIN, C.-Y., CHEN, C.-C., HUANG, H.-H., AND CHEN, H.-H. Argument-based sentiment analysis on forward-looking statements. In *Findings of the Association for Computational Linguistics ACL 2024* (Bangkok, Thailand and virtual meeting, Aug. 2024), L.-W. Ku, A. Martins, and V. Srikumar, Eds., Association for Computational Linguistics, pp. 13804–13815.
18. LIU, Y.-T., CHEN, C.-C., HUANG, H.-H., AND CHEN, H.-H. News-driven price movement forecasting with label-prior graph attention. In *Companion Proceedings of the ACM on Web Conference 2024* (2024), pp. 569–572.
19. MORIO, G., EGAWA, R., AND FUJITA, K. Revealing and predicting online persuasion strategy with elementary units. In *Proceedings of the 2019 Conference on Empirical Methods in Natural Language Processing and the 9th International Joint Conference on Natural Language Processing (EMNLP-IJCNLP)* (Hong Kong, China, Nov. 2019), Association for Computational Linguistics, pp. 6274–6279.
20. NISHINO, S., WASHIDA, Y., ISHIGAKI, T., WASHINO, S., IGARASHI, H., MURAI, A., AND NAGAI, Y. Validation of a foresight support system to imagine an uncertain future:- effectiveness testing through scenario planning workshops. *IIAI Letters on Informatics and Interdisciplinary Research 3* (2023).

21. SOHRABI, S., RIABOV, A., KATZ, M., AND UDREA, O. An ai planning solution to scenario generation for enterprise risk management. In *Proceedings of the AAAI Conference on Artificial Intelligence* (2018), vol. 32.
22. TANG, T.-H., CHEN, C.-C., HUANG, H.-H., AND CHEN, H.-H. Retrieving implicit information for stock movement prediction. In *Proceedings of the 44th International ACM SIGIR Conference on Research and Development in Information Retrieval* (2021), pp. 2010–2014.
23. ZONG, S., RITTER, A., AND HOVY, E. Measuring forecasting skill from text. In *Proceedings of the Fifty-Eighth Annual Meeting of the Association for Computational Linguistics* (Online, July 2020), Association for Computational Linguistics, pp. 5317–5331.

Open Access This chapter is licensed under the terms of the Creative Commons Attribution 4.0 International License (http://creativecommons.org/licenses/by/4.0/), which permits use, sharing, adaptation, distribution and reproduction in any medium or format, as long as you give appropriate credit to the original author(s) and the source, provide a link to the Creative Commons license and indicate if changes were made.

The images or other third party material in this chapter are included in the chapter's Creative Commons license, unless indicated otherwise in a credit line to the material. If material is not included in the chapter's Creative Commons license and your intended use is not permitted by statutory regulation or exceeds the permitted use, you will need to obtain permission directly from the copyright holder.

Chapter 3
Single-Agent/Model Design

In this chapter, we focus on single-agent/model design. Section 3.1 begins with a discussion comparing end-to-end learning and feature engineering in the modern era. This section primarily explores the value of manual annotation and how to incorporate human insights into models to enhance the performance of downstream tasks. Beyond learning from humans and experts, the ability to learn from historical events and documents is also crucial for agents. Accordingly, in Sect. 3.2, we examine how a retrieval-augmented approach can be used to generate multiple question candidates based on speech transcripts and to produce answer templates in professional presentation settings. Taking this further, instead of reading the retrieved documents each time, Sect. 3.3 explores whether models can internalize new knowledge through the model editing approach. Finally, we conclude and provide remarks in Sect. 3.4.

3.1 Learning from Human Insights

The evolution of AI has been marked by many transformative milestones. These advancements have reshaped traditional paradigms in NLP, particularly in how we approach task-specific training and annotation. Before the emergence of LLMs, the creation of annotated datasets was fundamental for training AI models, enabling them to achieve high accuracy on specific tasks [6, 18, 24, 37]. This approach, rooted in the principle of supervised learning, emphasized the importance of human expertise in curating high-quality datasets for model training.

The introduction of LLMs has challenged this paradigm by showcasing remarkable capabilities through few-shot and zero-shot learning. These models can generate fluent and coherent responses to various tasks by leveraging massive pre-training datasets, often reducing or eliminating the need for manual annotations [3, 44, 46]. Despite these advancements, questions remain about the extent to which human annotations are still necessary in the age of LLMs. This has sparked an ongoing debate within the NLP community: Can prompt-based LLMs entirely replace the traditional reliance on human-curated datasets? Or does the human touch retain its relevance in fine-tuning and extending the applicability of AI models?

To address these questions, researchers have undertaken comparative analyses to evaluate the performance of LLMs and PLMs trained on annotated datasets. This section will discuss the interplay between human insights and AI, exploring how annotated data continues to play a pivotal role in advancing both LLMs and PLMs. By examining experimental results and methodologies, we aim to illuminate the ongoing value of human annotation and its role in bridging performance gaps in complex, nuanced tasks. Additionally, we survey recent innovations, such as pre-finetuning techniques [2], that integrate human-derived insights to enhance the adaptability of AI models. Through this discussion, we argue that human expertise remains indispensable for imbuing models with the capacity to address novel challenges and incorporate domain-specific knowledge effectively.

Building on the discussion from the previous chapter, we demonstrate the value of human annotation with the results shown in Table 3.1. We conducted experiments with several PLMs and compared their best performance to that of zero-shot or few-shot LLMs [31]. As shown, the performance of all tasks decreases, especially for the new task of impact duration estimation, which has rarely been explored in previous work. These results support our claim that supervised AI models can perform better than prompt-based LLMs. On the basis of these findings, we suggest that annotation efforts remain necessary for models to learn specific tasks effectively in this era. However, while LLMs do not perform as well, their generalizability is still remarkable. For example, in argument-based sentiment analysis tasks, the performance gap is quite small, even when LLMs have not been trained with annotated data.

In line with this reasoning, we further discuss how human insights can be leveraged to improve the performance of both LLMs and PLMs. We apply the concept of impact duration to make additional annotations on social media threads [17]. Unlike the observations in Table 2.2, which are based on professional analysts' reports, social media users often provide very short-term views. Specifically, 24.17% of analyses are shared with reference to market movements within one week. The results in Table 3.1 show that even LLMs, despite being pre-trained with extensive amounts of data, struggle with the impact duration inference task. Therefore, we propose exploring the pre-finetuning scheme [2] to enhance the model's ability to utilize the

Table 3.1 Performance difference compared with the best PLM (Accuracy) [31]

LLM	Setting	Argument unit	Argument-based sentiment analysis	Impact duration
GPT-3.5	Zero-Shot	−0.182	−0.024	−0.237
GPT-3.5	Few-Shot	−0.151	−0.014	−0.215
GPT-4	Few-Shot	−0.131	−0.010	−0.207

3.1 Learning from Human Insights

Table 3.2 Performance improvement after pre-finetuning [17]

	Accuracy	F1
BERT-Chinese	4.81	4.59
Multilingual-BERT	4.54	4.68
Chinese-BERT	1.36	1.03
Mengzi-BERT	4.33	4.39
Mengzi-BERT-Fin	6.57	6.71

estimated impact duration. The primary concept of pre-finetuning is to introduce an additional learning stage between pre-training and fine-tuning. We use PLMs as baselines and adopt stock movement prediction [56] as the target fine-tuning task, with impact duration inference serving as the pre-finetuning task in our approach [17].

We use BERT-Chinese [20], Multilingual-BERT [20], Chinese-BERT [19], Mengzi-BERT [53], and Mengzi-BERT-Fin [53] in our experiments. Table 3.2 shows the performance improvement after adopting the impact duration pre-finetuning task. Regardless of the PLM used, performance improves after learning the impact duration ability. This implies that when the task is novel, adopting a pre-finetuning scheme can enhance the model's ability to adapt to new aspects, thereby facilitating its application to downstream tasks. Moreover, these results clarify the value of human insight in training models to consider specific (novel) concepts when dealing with a given task. While models have already demonstrated impressive performance in understanding and generating natural language through next-token prediction and masked word prediction, creativity and the identification of new features remain challenges. Therefore, if experts wish to inject novel features (concepts) into the models, pre-finetuning would be a suitable approach. We argue that pre-finetuning can be viewed as a type of feature engineering when designing a single agent or model.

One further piece of evidence is that numbers are crucial for financial document understanding, as introduced in Chap. 5 of our previous book [12]. There are many fine-grained meanings attached to numbers in financial documents [14, 15], and their associated entities are also important [10]. Following this line of thought, we further experiment on whether numeral-aware pre-finetuning can help improve a model's ability for downstream tasks [40]. We pre-finetune RoBERTa-large for a numeral attachment task and then use the pre-finetuned model for volatility forecasting tasks [13, 38, 51]. The performance (mean squared error) improvement after pre-finetuning is 0.030, 0.033, and 0.001 for 3, 7, and 15-day volatility forecasting, respectively. This suggests that the current PLMs may not be able to focus on key information typically used by experts for downstream tasks when only pre-trained with traditional settings.

Human insights are useful not only for PLMs but also for LLMs. Based on the premise that numbers are significant for understanding financial documents, we explore whether LLMs are already aware of the numbers when performing sentiment analysis on number-rich tweets [11]. We experiment with four LLMs: PaLM

Table 3.3 Performance improvement in sentiment analysis compared with simple prompting [16]

	CoT	CoT + Hint
PaLM	−1.75	−0.46
Gemini	3.68	6.19
GPT-3.5	0.56	11.25
GPT-4	−4.66	3.85

2 [5], Gemini Pro,[1] GPT-3.5, and GPT-4 [1]. Given that prompting is the simplest way to test LLMs' abilities, we explore three prompting strategies. The first strategy is to ask LLMs to predict the sentiment, and the second is to ask LLMs to perform chain-of-thought [49], providing step-by-step analysis before making a prediction. The third strategy provides expert hints in the prompt, asking models to analyze the numbers in the given tweet [16].

Table 3.3 presents the experimental results, showing the improvement in the micro-F1 score when compared with the simple prompting strategy. First, the results indicate that using expert hints leads to significant performance gains in three out of four LLMs. While the performance slightly decreases when using PaLM, the drop is not significant. This demonstrates that human insights are still worth exploring and remain important when using LLMs. On the other hand, while chain-of-thought prompting improves performance for Gemini and GPT-3.5, it decreases performance for PaLM and GPT-4. This further reveals that performance becomes more stable when expert hints are adopted for sentiment analysis.

To conclude this section, we present several examples highlighting the importance of annotated data and provide guidance on how to inject human insights into PLMs and LLMs. The intent is to demonstrate that feature engineering remains helpful when using PLMs and LLMs. In the near future, even though AI agents may suggest creative ideas [42], human insights and experience can still be employed to guide these agents in adopting both AI-based ideas and human-generated insights.

3.2 Retrieval-Augmented Generation

Recent advances in LLMs have demonstrated impressive capabilities in generating coherent and contextually relevant text. However, their performance often depends on the quality and scope of the input context. To overcome the inherent limitations of relying solely on pre-trained knowledge, RAG has emerged as a powerful paradigm. By retrieving external documents or knowledge, RAG enriches the input to generative models, thereby enabling more informed and accurate outputs.

---

[1] https://deepmind.google/technologies/gemini/.

3.2 Retrieval-Augmented Generation

This technique has been widely applied in tasks such as open-domain question answering [30, 43, 47], knowledge-grounded dialogue systems [4, 55], and document summarization [26, 48].

RAG combines the strengths of information retrieval and generative modeling, allowing systems to dynamically incorporate relevant external information during inference. Research in this domain has explored diverse approaches, from unsupervised retrieval techniques like BM25 [39] to neural retrievers such as dense passage retrieval (DPR) [29]. Despite its success, RAG's application to specific real-world scenarios, such as interactive and adaptive tasks in professional settings, remains underexplored. In this section, we extend the discussion of RAG by addressing its utility in a unique yet practical task: assisting presenters in preparing for question-answering sessions during oral presentations. Drawing from the financial domain, we employ earnings call transcripts as a testbed for our experiments. Earnings calls are regular meetings where company managers deliver prepared remarks and respond to questions from professional analysts. Unlike traditional reading comprehension tasks, where answers are confined to a provided context, the questions posed during these calls often require inference, synthesis, or extrapolation beyond the immediate content. This distinctive challenge positions the task as an open-field question generation problem, which demands novel retrieval and generation strategies to enhance performance.

Building upon foundational work in question generation and knowledge augmentation, we explore methods to refine and enrich model inputs using retrieved key paragraphs and causal KGs. By combining retriever-driven paragraph selection with causal reasoning frameworks, we aim to elevate the utility of RAG in this demanding application. The following sections detail our methodologies, experimental results, and insights gained from integrating these components into a cohesive system for oral presentation preparation.

Generating questions for reading comprehension tasks is the most common question generation task [35]. Although considering potential questions from the audience is a common task for presenters, there has been limited discussion on this type of task setting. The primary difference between reading comprehension-type question generation and oral presentation question generation is that the answer to the former is either provided or exists within the article. However, when generating questions for oral presentations, it becomes an open-field question generation task. Professional analysts rarely ask managers to simply repeat or reframe sentences from the presentation: they are more interested in probing for deeper details or exploring entirely new topics. This makes the task more challenging than reading comprehension-type question generation. Furthermore, rather than generating a single question, models are expected to generate multiple questions for the same presentation script, a task known as multiple question generation (MQG) [27].

To address this task, we first propose training a retriever to identify key paragraphs that may be relevant to the professional analysts' questions, followed by training models to generate questions based on these key paragraphs. We employ LongFormer [8], LongT5 [23], and FROST [34] for our pilot explorations. Compared with using the previous N tokens as model input, using the extracted key paragraphs for MQG

Table 3.4 Performance improvement compared to the model without a retriever

Retriver	BLEU-4	ROUGE-L	BERTScore
Random	1.121	5.654	1.110
BM25	1.107	6.298	2.158
Co-trained retriver	1.471	7.258	4.700

improves ROUGE-L scores by 4.89, 3.95, and 3.85 for each model, respectively. These results demonstrate the importance of retrieval in the MQG task.

Taking one step further, we attempt to co-train the retriever with the generator [28]. Specifically, we prompt the LLM to determine whether a given paragraph is highly, partially, or not related to a given question, subsequently using the top-k paragraphs to generate questions. We explore this setting with Alpaca-Lora-7B [25, 45]. For comparison, models without a retriever and those utilizing either a random paragraph or BM25 as a retriever are considered. Table 3.4 presents the results. First, the use of retrievers improves performance across all evaluation metrics, thus highlighting the significance of incorporating a retriever for this task. Second, the co-trained retriever outperforms both randomly chosen paragraphs and BM25 as retrievers.

In summary, while LLMs are effective in generating text, providing additional assistance and references, as demonstrated in the MQG results, can significantly enhance performance.

In addition to retrieving key paragraphs for refining model input, acquiring additional knowledge is also a common approach. Given that answering questions from professional analysts requires some causal inference ability, we utilize causal knowledge graphs (KGs) to enhance the knowledge of LLMs when generating responses. Here, we aim to recap the notion of forward-looking argument mining discussed in Chap. 2, and further connect this concept with causal knowledge graph construction. As mentioned, premises or scenarios are used to support claims. By segmenting an analysis into argument units, an opinion graph can be formed, which can serve as a foundation for the causal KG. For example, the entities within the premise and the claim can become the nodes of the causal KG, with directed edges from premise to claim, such as "sales" influencing "revenue." Following this line of reasoning, we utilize Equity-AMSA [31] to construct a causal KG.

As in the previous experiments, we use earnings call transcripts as data [41], modifying the task to examine template generation. We explore GPT-3.5, Gemini-Pro, and LLaMA-3 8B in the experiment, evaluating performance based on whether the financial terms in the manager's answer also appear in the generated answer template. Specifically, we assess whether the models can suggest answering directions as correctly as managers do. In our approach, we follow the design of previous work [7], retrieving relevant causal KGs based on the given question and incorporating the KG information into the prompt for the LLM to reference. Table 3.5 displays the performance improvement compared to the setting without KG. In addition to the causal KG we constructed (Equity-AMSA-KG), we compare the performance with

3.3 Model Editing

Table 3.5 Results after adding KG [41]. The numbers in parentheses represent the performance improvement compared to using LLM directly without KG

KG	–	FinCaKG-FR [50]	FinCaKG-ECT	Equity-AMSA-KG
Source		10-K report	Earnings calls	Research report
GPT-3.5	22.35	21.62 (−0.73)	22.38 (0.03)	24.06 (1.71)
Gemini Pro	15.47	11.83 (−3.64)	13.83 (−1.64)	15.91 (0.44)
LLaMA-3 8B	19.53	19.98 (0.45)	20.52 (0.99)	19.08 (−0.45)

two other causal KGs that were constructed using different sources [50]. The results indicate that KGs contribute to performance enhancement, though no specific KG works universally across all LLMs.

In conclusion, we demonstrate the potential of retrieval in a real-world scenario within the financial domain: oral presentation preparation. The retriever can refine model input and also retrieve additional knowledge to enrich the input.

3.3 Model Editing

Maintaining LLM relevance and accuracy over time has become more and more critical, leading to an increased focus on strategies for updating models to integrate new knowledge or correct errors without retraining them from scratch. Among these strategies, model editing has emerged as a promising approach. Unlike pre-finetuning or RAG, which involve either large-scale tuning of model parameters or reliance on external documents, model editing focuses on modifying a targeted subset of parameters to achieve specific updates efficiently. This section provides an overview of the concept of model editing, its motivations, and key methodologies, before providing our analysis of the effects of question types on editing outcomes.

Several recent works have explored various facets of model editing, including techniques to localize updates within a model [9, 21, 32, 33, 52, 54]. These studies have demonstrated that model editing offers a flexible alternative to traditional fine-tuning, enabling users to inject or modify knowledge while preserving the model's overall capabilities. However, this flexibility is not without challenges. Model editing can lead to unintended side effects, such as performance degradation on unrelated tasks or inconsistencies across different queries [22]. Addressing these challenges is critical to ensuring the reliability and robustness of edited models.

A key area of interest in model editing research involves evaluating the impacts of different editing techniques across diverse contexts and datasets. While prior work has focused on the efficiency and accuracy of various methods, limited attention has been given to how the nature of the queries being edited–such as the type of question–affects the outcomes. This gap in understanding is particularly relevant, because language models are increasingly used in diverse real-world applications where question types and contexts can vary widely.

In this section, we build upon prior research and introduce a novel perspective by analyzing model editing performance across different question types. We employ two popular editing techniques, fine-tuning and the MEMIT method [32], and conduct experiments on two language models, GPT-2 and LLaMA-2, to investigate whether model size influences the results. Fine-tuning involves retraining a language model on new data to update its knowledge, modifying the model's weights globally. MEMIT, in contrast, is designed to efficiently insert or modify specific knowledge within large language models. By adjusting select internal layers, MEMIT enables precise edits without requiring full model retraining. By examining both the immediate and cumulative side effects of edits, we aim to provide insights into the trade-offs and limitations of different editing approaches. Figure 3.1 presents the experimental results. Although side effects occur regardless of the question type, the degree of performance difference varies among different types. When using the fine-tuning approach, "why" questions lead to a smaller performance difference, irrespective of the model size. In contrast, when adopting the MEMIT approach, GPT-2 drops to a similar level regardless of the question type, whereas LLaMA-2 drops to 0.50 after the first editing, with the timing of subsequent drops differing among question types. For "why" questions, there is only one significant drop, while for "when" and "where" questions, the performance remains consistent after multiple edits. However, for other question types, a significant drop occurs after a certain number of edits. This indicates that the MEMIT editing approach may have markedly different side effects on LLMs. Such differences are not observed when using the fine-tuning approach,

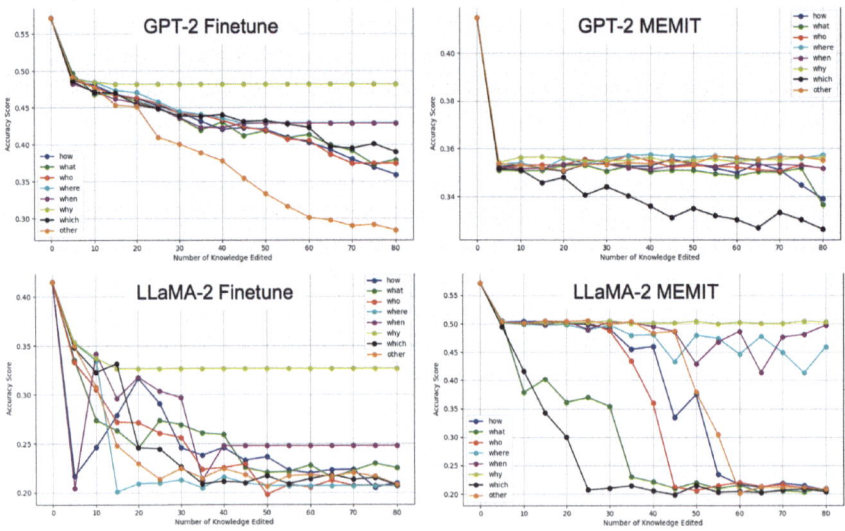

Fig. 3.1 Question-type-based side effect analysis [36]

suggesting that fine-tuning leads to a consistent performance level after multiple edits. Furthermore, the trends observed with GPT-2 (small model) differ from those with LLaMA-2 (LLM), implying that the same editing method may not be effective for both types of models.

In summary, we have highlighted an alternative popular method for updating LMs and explored a novel research question regarding the side effects of different question types. Our results reveal indicate that findings from small models cannot be directly applied to large models, and the side effects of different editing methods also differ significantly.

3.4 Summary

In this chapter, we delved into strategies for designing a single agent/model and emphasized the integration of human insights, retrieval-augmented generation, and model editing to enhance performance.

We began by exploring the enduring value of human annotation in the age of LLMs. Despite the capabilities of LLMs to generate fluent responses through prompting, our experiments revealed that supervised models trained on human-annotated data often outperform prompt-based LLMs, especially in specialized tasks like forward-looking argument mining. For example, tasks such as impact duration estimation–a novel and less-explored area–show significant performance drops when relying solely on LLMs without annotated data.

We further discussed the concept of pre-finetuning, an intermediate training stage between pre-training and fine-tuning. By pre-finetuning models on tasks that incorporate human insights, such as impact duration inference and numeral attachment, we observed notable performance improvements in downstream applications like stock movement prediction and volatility forecasting. This approach demonstrates that human expertise and novel features can be effectively integrated into models, and enhance their adaptability and performance.

Additionally, we examined how providing expert hints in prompts can benefit LLMs. Through experiments involving sentiment analysis of number-rich tweets using models like GPT-3.5 and GPT-4, we found that incorporating human insights into prompts leads to significant performance gains. This suggests that even advanced LLMs can benefit from human-guided feature engineering.

Next, we addressed the importance of retrieving relevant information to augment model generation capabilities. We presented a practical application where models assist presenters in preparing for question-and-answer sessions by generating potential questions based on earnings call transcripts. Unlike traditional question generation tasks, this scenario involves open-field questions that may delve deeper or introduce new topics not explicitly covered in the text.

By training a retriever to identify key paragraphs and co-training it with a question generator, we achieved improved performance in multiple question generation tasks. Models like LongFormer and LongT5 benefit from this approach, as evidenced by significant increases in ROUGE-L scores. Furthermore, integrating external knowledge through causal KGs enhances the models' ability to suggest accurate answering directions and templates, demonstrating the efficacy of retrieval-augmented generation in complex, real-world tasks.

Finally, we explored model editing as a means to update language models with new knowledge by modifying only a few parameters. This approach allows models to internalize updates without the need for extensive retraining or repeatedly supplying external documents.

Our analysis focused on the side effects of model editing across different question types using methods like fine-tuning and MEMIT. We discovered that the impact of model editing varies with both the question type and the size of the model. For example, "why" questions tend to experience smaller performance drops when using fine-tuning, regardless of model size. In contrast, when applying MEMIT to larger models like LLaMA-2, we observed significant performance drops that vary depending on the question type. These findings indicate that the effectiveness and side effects of model editing methods are not uniform across different models and question types, highlighting the need for careful consideration in their application.

Overall, this chapter highlights the ongoing importance of human insights and feature engineering in AI model development. Despite the advancements in LLMs, human expertise remains crucial for training models to perform specific, nuanced tasks effectively. Techniques like pre-finetuning, retrieval-augmented generation, and model editing serve as valuable tools for integrating human knowledge and addressing the limitations of models trained solely on large datasets. As AI agents continue to evolve, the collaboration between human insights and AI capabilities will be essential in achieving optimal performance in specialized domains.

References

1. ACHIAM, J., ADLER, S., AGARWAL, S., AHMAD, L., AKKAYA, I., ALEMAN, F. L., ALMEIDA, D., ALTENSCHMIDT, J., ALTMAN, S., ANADKAT, S., ET AL. Gpt-4 technical report. *ar*Xiv preprint arXiv:2303.08774 (2023).
2. AGHAJANYAN, A., GUPTA, A., SHRIVASTAVA, A., CHEN, X., ZETTLEMOYER, L., AND GUPTA, S. Muppet: Massive multi-task representations with pre-finetuning. In *Proceedings of the 2021 Conference on Empirical Methods in Natural Language Processing* (Online and Punta Cana, Dominican Republic, Nov. 2021), M.-F. Moens, X. Huang, L. Specia, and S. W.-t. Yih, Eds., Association for Computational Linguistics, pp. 5799–5811.
3. AGUDA, T. D., SIDDAGANGAPPA, S., KOCHKINA, E., KAUR, S., WANG, D., AND SMILEY, C. Large language models as financial data annotators: A study on effectiveness and efficiency. In *Proceedings of the 2024 Joint International Conference on Computational Linguistics, Language Resources and Evaluation (LREC-COLING 2024)* (Torino, Italia, May 2024), N. Calzolari, M.-Y. Kan, V. Hoste, A. Lenci, S. Sakti, and N. Xue, Eds., ELRA and ICCL, pp. 10124–10145.
4. ALGHISI, S., RIZZOLI, M., ROCCABRUNA, G., MOUSAVI, S. M., AND RICCARDI, G. Should we fine-tune or RAG? evaluating different techniques to adapt LLMs for dialogue. In *Proceedings of the 17th International Natural Language Generation Conference* (Tokyo, Japan,

Sept. 2024), S. Mahamood, N. L. Minh, and D. Ippolito, Eds., Association for Computational Linguistics, pp. 180–197.
5. ANIL, R., DAI, A. M., FIRAT, O., JOHNSON, M., LEPIKHIN, D., PASSOS, A., SHAKERI, S., TAROPA, E., BAILEY, P., CHEN, Z., ET AL. Palm 2 technical report. arXiv preprint arXiv:2305.10403 (2023).
6. ATANASOVA, P., BARRON-CEDENO, A., ELSAYED, T., SUWAILEH, R., ZAGHOUANI, W., KYUCHUKOV, S., MARTINO, G. D. S., AND NAKOV, P. Overview of the clef-2018 checkthat! lab on automatic identification and verification of political claims. task 1: Checkworthiness, 2018.
7. BAEK, J., AJI, A. F., AND SAFFARI, A. Knowledge-augmented language model prompting for zero-shot knowledge graph question answering. In *P*roceedings of the 1st Workshop on Natural Language Reasoning and Structured Explanations (NLRSE) (Toronto, Canada, June 2023), B. Dalvi Mishra, G. Durrett, P. Jansen, D. Neves Ribeiro, and J. Wei, Eds., Association for Computational Linguistics, pp. 78–106.
8. BELTAGY, I., PETERS, M. E., AND COHAN, A. Longformer: The long-document transformer. arXiv preprint arXiv:2004.05150 (2020).
9. CAO, N. D., AZIZ, W., AND TITOV, I. Editing factual knowledge in language models. In *C*onference on Empirical Methods in Natural Language Processing (2021).
10. CHEN, C.-C., HUANG, H.-H., AND CHEN, H.-H. Numeral attachment with auxiliary tasks. In *P*roceedings of the Forty-Second International ACM SIGIR Conference on Research and Development in Information Retrieval (2019), pp. 1161–1164.
11. CHEN, C.-C., HUANG, H.-H., AND CHEN, H.-H. Issues and perspectives from 10,000 annotated financial social media data. In *P*roceedings of the Twelfth Language Resources and Evaluation Conference (Marseille, France, May 2020), N. Calzolari, F. Béchet, P. Blache, K. Choukri, C. Cieri, T. Declerck, S. Goggi, H. Isahara, B. Maegaard, J. Mariani, H. Mazo, A. Moreno, J. Odijk, and S. Piperidis, Eds., European Language Resources Association, pp. 6106–6110.
12. CHEN, C.-C., HUANG, H.-H., AND CHEN, H.-H. *F*rom opinion mining to financial argument mining. Springer Nature, 2021.
13. CHEN, C.-C., HUANG, H.-H., HUANG, Y.-L., AND CHEN, H.-H. Distilling numeral information for volatility forecasting. In *P*roceedings of the 30th ACM International Conference on Information & Knowledge Management (New York, NY, USA, 2021), Association for Computing Machinery, p. 2920–2924.
14. CHEN, C.-C., HUANG, H.-H., SHIUE, Y.-T., AND CHEN, H.-H. Numeral understanding in financial tweets for fine-grained crowd-based forecasting. In 2018 IEEE/WIC/ACM International Conference on Web Intelligence (WI) (2018), IEEE, pp. 136–143.
15. CHEN, C.-C., HUANG, H.-H., TAKAMURA, H., AND CHEN, H.-H. Overview of the NTCIR-14 FinNum Task: Fine-grained numeral understanding in financial social media data. In *P*roceedings of the Fourteenth NTCIR Conference on Evaluation of Information Access Technologies (2019), pp. 19–27.
16. CHEN, C.-C., TAKAMURA, H., KOBAYASHI, I., AND MIYAO, Y. Enhancing financial sentiment analysis with expert-designed hint. In arXiv (2024).
17. CHIU, C.-J., CHEN, C.-C., HUANG, H.-H., AND CHEN, H.-H. Pre-finetuning with impact duration awareness for stock movement prediction. In *C*ompanion Proceedings of the Web Conference 2025 (WWW'25) (2025).
18. CORTIS, K., FREITAS, A., DAUDERT, T., HUERLIMANN, M., ZARROUK, M., HANDSCHUH, S., AND DAVIS, B. SemEval-2017 task 5: Fine-grained sentiment analysis on financial microblogs and news. In *P*roceedings of the 11th International Workshop on Semantic Evaluation (SemEval-2017) (Vancouver, Canada, Aug. 2017), S. Bethard, M. Carpuat, M. Apidianaki, S. M. Mohammad, D. Cer, and D. Jurgens, Eds., Association for Computational Linguistics, pp. 519–535.
19. CUI, Y., CHE, W., LIU, T., QIN, B., AND YANG, Z. Pre-training with whole word masking for chinese bert. *I*EEE/ACM Transactions on Audio, Speech, and Language Processing 29 (2021), 3504–3514.

20. DEVLIN, J., CHANG, M.-W., LEE, K., AND TOUTANOVA, K. BERT: Pre-training of deep bidirectional transformers for language understanding. In *Proceedings of the 2019 Conference of the North American Chapter of the Association for Computational Linguistics: Human Language Technologies, Volume 1 (Long and Short Papers)* (Minneapolis, Minnesota, June 2019), Association for Computational Linguistics, pp. 4171–4186.
21. GEVA, M., SCHUSTER, R., BERANT, J., AND LEVY, O. Transformer feed-forward layers are key-value memories. In *Proceedings of the 2021 Conference on Empirical Methods in Natural Language Processing* (2021), pp. 5484–5495.
22. GU, J.-C., XU, H.-X., MA, J.-Y., LU, P., LING, Z.-H., CHANG, K.-W., AND PENG, N. Model editing can hurt general abilities of large language models. In *EMNLP* (2024).
23. GUO, M., AINSLIE, J., UTHUS, D., ONTANON, S., NI, J., SUNG, Y.-H., AND YANG, Y. LongT5: Efficient text-to-text transformer for long sequences. In *Findings of the Association for Computational Linguistics: NAACL 2022* (Seattle, United States, July 2022), Association for Computational Linguistics, pp. 724–736.
24. HASSAN, N., LI, C., AND TREMAYNE, M. Detecting check-worthy factual claims in presidential debates. In *Proceedings of the 24th acm international on conference on information and knowledge management* (2015), pp. 1835–1838.
25. HU, E. J., WALLIS, P., ALLEN-ZHU, Z., LI, Y., WANG, S., WANG, L., CHEN, W., ET AL. Lora: Low-rank adaptation of large language models. In *International Conference on Learning Representations*.
26. JI, Y., LI, Z., MENG, R., SIVARAJKUMAR, S., WANG, Y., YU, Z., JI, H., HAN, Y., ZENG, H., AND HE, D. RAG-RLRC-LaySum at BioLaySumm: Integrating retrieval-augmented generation and readability control for layman summarization of biomedical texts. In *Proceedings of the 23rd Workshop on Biomedical Natural Language Processing* (Bangkok, Thailand, Aug. 2024), D. Demner-Fushman, S. Ananiadou, M. Miwa, K. Roberts, and J. Tsujii, Eds., Association for Computational Linguistics, pp. 810–817.
27. JUAN, Y., CHEN, C.-C., HUANG, H.-H., AND CHEN, H.-H. Generating multiple questions from presentation transcripts: A pilot study on earnings conference calls. In *Proceedings of the 16th International Natural Language Generation Conference* (Prague, Czechia, Sept. 2023), C. M. Keet, H.-Y. Lee, and S. Zarrieß, Eds., Association for Computational Linguistics, pp. 449–454.
28. JUAN, Y., CHEN, C.-C., HUANG, H.-H., AND CHEN, H.-H. Co-trained retriever-generator framework for question generation in earnings calls. In *Companion Proceedings of the Web Conference 2025* (2025).
29. KARPUKHIN, V., OGUZ, B., MIN, S., LEWIS, P., WU, L., EDUNOV, S., CHEN, D., AND YIH, W.-T. Dense passage retrieval for open-domain question answering. In *Proceedings of the 2020 Conference on Empirical Methods in Natural Language Processing (EMNLP)* (Online, Nov. 2020), B. Webber, T. Cohn, Y. He, and Y. Liu, Eds., Association for Computational Linguistics, pp. 6769–6781.
30. KIM, K., AND LEE, J.-Y. RE-RAG: Improving open-domain QA performance and interpretability with relevance estimator in retrieval-augmented generation. In *Proceedings of the 2024 Conference on Empirical Methods in Natural Language Processing* (Miami, Florida, USA, Nov. 2024), Y. Al-Onaizan, M. Bansal, and Y.-N. Chen, Eds., Association for Computational Linguistics, pp. 22149–22161.
31. LIN, C.-Y., CHEN, C.-C., HUANG, H.-H., AND CHEN, H.-H. Argument-based sentiment analysis on forward-looking statements. In *Findings of the Association for Computational Linguistics ACL 2024* (Bangkok, Thailand and virtual meeting, Aug. 2024), L.-W. Ku, A. Martins, and V. Srikumar, Eds., Association for Computational Linguistics, pp. 13804–13815.
32. MENG, K., SHARMA, A. S., ANDONIAN, A. J., BELINKOV, Y., AND BAU, D. Mass-editing memory in a transformer. In *The Eleventh International Conference on Learning Representations*.
33. MITCHELL, E., LIN, C., BOSSELUT, A., MANNING, C. D., AND FINN, C. Memory-based model editing at scale. In *International Conference on Machine Learning* (2022), PMLR, pp. 15817–15831.

References

34. NARAYAN, S., ZHAO, Y., MAYNEZ, J., SIMÕES, G., NIKOLAEV, V., AND MCDONALD, R. Planning with learned entity prompts for abstractive summarization. *T*ransactions of the Association for Computational Linguistics 9 (2021), 1475–1492.
35. PAN, L., LEI, W., CHUA, T.-S., AND KAN, M.-Y. Recent advances in neural question generation. *ar*Xiv e-prints (2019), arXiv–1905.
36. PAN, T.-H., CHEN, C.-C., HUANG, H.-H., AND CHEN, H.-H. "why" has the least side effect on model editing. In *ar*Xiv (2024).
37. PATWARI, A., GOLDWASSER, D., AND BAGCHI, S. Tathya: A multi-classifier system for detecting check-worthy statements in political debates. In *P*roceedings of the 2017 ACM on Conference on Information and Knowledge Management (2017), pp. 2259–2262.
38. QIN, Y., AND YANG, Y. What you say and how you say it matters: Predicting stock volatility using verbal and vocal cues. In *P*roceedings of the Fifty-Seventh Annual Meeting of the Association for Computational Linguistics (Florence, Italy, July 2019), Association for Computational Linguistics, pp. 390–401.
39. ROBERTSON, S. E., AND WALKER, S. Some simple effective approximations to the 2-poisson model for probabilistic weighted retrieval. In *SIGIR'94: Proceedings of the Seventeenth Annual International ACM-SIGIR Conference on Research and Development in Information Retrieval*, organised by Dublin City University (1994), Springer, pp. 232–241.
40. SHI, M.-X., CHEN, C.-C., HUANG, H.-H., AND CHEN, H.-H. Enhancing volatility forecasting in financial markets: A general numeral attachment dataset for understanding earnings calls. In *P*roceedings of the 13th International Joint Conference on Natural Language Processing and the 3rd Conference of the Asia-Pacific Chapter of the Association for Computational Linguistics (Volume 2: Short Papers) (Nusa Dua, Bali, Nov. 2023), J. C. Park, Y. Arase, B. Hu, W. Lu, D. Wijaya, A. Purwarianti, and A. A. Krisnadhi, Eds., Association for Computational Linguistics, pp. 37–42.
41. SHIH, Y.-Y., XU, Z., TAKAMURA, H., CHEN, Y.-N., AND CHEN, C.-C. Rehearsing answers to probable questions with perspective-taking. In *ar*Xiv (2024).
42. SI, C., YANG, D., AND HASHIMOTO, T. Can llms generate novel research ideas? a large-scale human study with 100+ nlp researchers. *ar*Xiv preprint arXiv:2409.04109 (2024).
43. SIRIWARDHANA, S., WEERASEKERA, R., WEN, E., KALUARACHCHI, T., RANA, R., AND NANAYAKKARA, S. Improving the domain adaptation of retrieval augmented generation (RAG) models for open domain question answering. *T*ransactions of the Association for Computational Linguistics 11 (2023), 1–17.
44. TAN, Z., LI, D., WANG, S., BEIGI, A., JIANG, B., BHATTACHARJEE, A., KARAMI, M., LI, J., CHENG, L., AND LIU, H. Large language models for data annotation and synthesis: A survey. In *P*roceedings of the 2024 Conference on Empirical Methods in Natural Language Processing (Miami, Florida, USA, Nov. 2024), Y. Al-Onaizan, M. Bansal, and Y.-N. Chen, Eds., Association for Computational Linguistics, pp. 930–957.
45. TAORI, R., GULRAJANI, I., ZHANG, T., DUBOIS, Y., LI, X., GUESTRIN, C., LIANG, P., AND HASHIMOTO, T. B. Stanford alpaca: An instruction-following llama model, 2023.
46. TSENG, Y.-M., CHEN, W.-L., CHEN, C.-C., AND CHEN, H.-H. Are expert-level language models expert-level annotators? In *ar*Xiv (2024).
47. WANG, Y., REN, R., LI, J., ZHAO, X., LIU, J., AND WEN, J.-R. REAR: A relevance-aware retrieval-augmented framework for open-domain question answering. In *P*roceedings of the 2024 Conference on Empirical Methods in Natural Language Processing (Miami, Florida, USA, Nov. 2024), Y. Al-Onaizan, M. Bansal, and Y.-N. Chen, Eds., Association for Computational Linguistics, pp. 5613–5626.
48. WANG, Z., TEO, S., OUYANG, J., XU, Y., AND SHI, W. M-RAG: Reinforcing large language model performance through retrieval-augmented generation with multiple partitions. In *P*roceedings of the 62nd Annual Meeting of the Association for Computational Linguistics (Volume 1: Long Papers) (Bangkok, Thailand, Aug. 2024), L.-W. Ku, A. Martins, and V. Srikumar, Eds., Association for Computational Linguistics, pp. 1966–1978.
49. WEI, J., WANG, X., SCHUURMANS, D., BOSMA, M., XIA, F., CHI, E., LE, Q. V., ZHOU, D., ET AL. Chain-of-thought prompting elicits reasoning in large language models. Advances in neural information processing systems 35 (2022), 24824–24837.

50. XU, Z., TAKAMURA, H., AND ICHISE, R. A framework to construct financial causality knowledge graph from text. In *2024 IEEE 18th International Conference on Semantic Computing (ICSC)* (2024), IEEE, pp. 57–64.
51. YANG, L., NG, T. L. J., SMYTH, B., AND DONG, R. Html: Hierarchical transformer-based multi-task learning for volatility prediction. In *Proceedings of The Web Conference 2020* (New York, NY, USA, 2020), WWW '20, Association for Computing Machinery, p. 441–451.
52. YAO, Y., WANG, P., TIAN, B., CHENG, S., LI, Z., DENG, S., CHEN, H., AND ZHANG, N. Editing large language models: Problems, methods, and opportunities. In *Conference on Empirical Methods in Natural Language Processing* (2023).
53. ZHANG, Z., ZHANG, H., CHEN, K., GUO, Y., HUA, J., WANG, Y., AND ZHOU, M. Mengzi: Towards lightweight yet ingenious pre-trained models for chinese. *arXiv preprint arXiv:2110.06696* (2021).
54. ZHU, C., LI, D., YU, F., ZAHEER, M., KUMAR, S., BHOJANAPALLI, S., AND RAWAT, A. S. Modifying memories in transformer models. In *International Conference on Machine Learning (ICML)* (2021), no. 2020.
55. ZHU, Z., LIAO, Y., XU, C., GUAN, Y., WANG, Y., AND WANG, Y. RA2FD: Distilling faithfulness into efficient dialogue systems. In *Proceedings of the 2024 Conference on Empirical Methods in Natural Language Processing* (Miami, Florida, USA, Nov. 2024), Y. Al-Onaizan, M. Bansal, and Y.-N. Chen, Eds., Association for Computational Linguistics, pp. 12304–12317.
56. ZOU, J., CAO, H., LIU, L., LIN, Y., ABBASNEJAD, E., AND SHI, J. Q. Astock: A new dataset and automated stock trading based on stock-specific news analyzing model. In *Proceedings of the Fourth Workshop on Financial Technology and Natural Language Processing* (2022).

Open Access This chapter is licensed under the terms of the Creative Commons Attribution 4.0 International License (http://creativecommons.org/licenses/by/4.0/), which permits use, sharing, adaptation, distribution and reproduction in any medium or format, as long as you give appropriate credit to the original author(s) and the source, provide a link to the Creative Commons license and indicate if changes were made.

The images or other third party material in this chapter are included in the chapter's Creative Commons license, unless indicated otherwise in a credit line to the material. If material is not included in the chapter's Creative Commons license and your intended use is not permitted by statutory regulation or exceeds the permitted use, you will need to obtain permission directly from the copyright holder.

Chapter 4
Multi-agent Interaction

Teamwork, or working collaboratively, has the potential to increase productivity, foster innovation, enhance problem-solving abilities, and improve skills among members. It is commonly practiced in professional settings. With the development of AI agents, there is an opportunity to enable agents to discuss with each other using natural language. After discussing the design of a single agent/model in the last chapter, this chapter explores the potential of teamwork among AI agents. As mentioned in Sect. 3.1, data annotation is one of the essential steps in AI research. Therefore, in Sect. 4.1, we explore whether it is possible to simulate the process of human annotation through multi-round discussions among agents. In addition to the classification task, we further discuss how to utilize a multi-agent interaction framework to improve generated results. In Sect. 4.2, we examine the improvement of decision-making by simulating hierarchical organization. In addition to traditional tasks, we discuss how to use multi-agents to simulate human behavior in Sect. 4.3. This section provides an early discussion of the trade-off of developing an unbiased agent. This chapter concludes in Sect. 4.4 with a summary.

4.1 Multi-round Discussion

In both human and AI-driven workflows, iterative refinement processes have been shown to enhance outcomes in tasks requiring complex reasoning or consensus-building. Multi-round discussion frameworks, inspired by human practices such as committee deliberations and collaborative annotation, have emerged as a promising approach in this context. These frameworks leverage iterative exchanges to address disagreements, reconcile diverse perspectives, and refine outputs. For data annotation and content generation tasks, such methodologies offer a structured way to simulate the advantages of collective reasoning and debate.

© The Author(s) 2025
C.-C. Chen and H. Takamura, *Agent AI for Finance*,
SpringerBriefs in Intelligent Systems,
https://doi.org/10.1007/978-3-031-94687-5_4

The growing reliance on LLMs to tackle these tasks underscores the importance of exploring mechanisms to improve their outputs beyond single-pass generation. While LLMs have demonstrated impressive capabilities, challenges such as inherent biases, limited domain-specific understanding, and the inability to reliably self-evaluate their outputs persist [14]. The potential of multi-round discussions lies in their ability to incorporate feedback, enhance interpretability, and achieve higher consensus quality, particularly in settings where direct human supervision may be limited.

When annotating data, disagreements among annotators are common. Majority voting is a typical method for determining the final label, and another approach is to gather annotators to discuss and reach a consensus. As discussed in Sect. 3.1, a single LLM may still not match the annotation quality achieved by humans [26], which raises the question of whether simulating the human annotation process could enhance quality. In this section, we share our experiences in adopting multi-round discussions for data annotation. We examine how these frameworks emulate human annotation practices, their effectiveness across diverse domains, and their limitations in novel or under-explored tasks. Additionally, we discuss their potential for generating complex, insight-driven analysis reports, highlighting both the opportunities and challenges these approaches present.

In our study [28], we experiment on six datasets spanning financial [17, 25], biomedical [8, 15], and legal [11, 13] domains. Three agents (GPT-4o, Gemini-1.5-pro, and Claude-3-opus) participate in the discussions. Specifically, each agent provides explanations and annotations based on the guideline in the first round. If disagreements arise among agents, their explanations are shared with one another for consideration in the second round. In the third round, the discussion history is shared for reference. If a consistent label is still not reached, majority voting is employed for evaluation. The accuracy improvements when using the multi-agent discussion framework over a single-agent framework are 3.9, 1.1, and 0.9% in the financial, biomedical, and legal datasets, respectively. These results indicate the potential of the multi-agent framework in a discussion setting for data annotation and suggest that simulating human workflows can lead to better results. While performance improves across domains, costs also increase significantly, indicating a trade-off between performance and cost.

Since previous tasks such as relation extraction and sentiment analysis have been widely studied, we here explore a relatively new task: multi-lingual ESG (Environmental, Social, and Governance) issue identification [7]. In this task, models must identify the most relevant ESG issue from 35 candidates for a given news article. Compared to the best-performing single-agent, the multi-agent discussion framework results in a performance drop of 1.0 and 1.4% for English and French datasets, respectively [27]. While this might suggest that multi-agent frameworks excel in well-established tasks but may not directly translate to improved outcomes in less-studied areas, alternative factors could also explain the observed performance gap. The complexity of the ESG domain, the novelty of the task, or challenges inherent to multi-lingual documents might influence the results. Therefore, further investigation is needed to disentangle these effects and better understand the limitations and potential of multi-agent approaches in emerging tasks.

4.1 Multi-round Discussion

However, although the multi-agent discussion framework cannot guarantee the best performance, it provides stable and reliable results. While pursuing optimal performance is the ultimate goal, having a method that consistently delivers a good performance is also worth exploring [4]. For example, we might need to explore a large number of LLMs to achieve the best performance, which may only suit a specific dataset. If the multi-agent discussion approach can consistently yield a good performance (not far from the best), users can easily establish a reasonable baseline for new tasks.

In addition to classification tasks, multi-agent discussion can also be used in generation tasks. For example, when discussing a new idea, having a meeting together would be a good method. When considering future scenarios, gathering experts together to perform scenario planning is common. One of the important characteristics of this setting is that participants are expected to have various backgrounds and to share diverse opinions. Therefore, we leverage the multi-agent discussion framework for analysis report generation. That is, instead of summarizing facts, agents are asked to generate insights as professional analysts.

When professional analysts write a report, they provide a brief summary of recent events and then propose some scenarios that may happen in the future. Some claims on earnings or price movement may also be provided in the report. Although previous studies have proposed various summarization challenges [21], the discussion on report or analysis generation is still in the early stage. Our pilot exploration focuses on generating reports after the earning calls [10]. The discussion framework asks multiple agents to take a look at different aspects (audio, company's past performance, etc) and then generate an analysis based on the given data. One writer agent will gather the entire analysis together to write a report. Compared to using a writer agent directly, including multiple aspects of opinions in the discussion framework can provide more insightful and useful information. Despite the positive results, the evaluation problem faced by most recent generation-related studies is also challenging in the report generation task. First, evaluating opinion is a subjective task, and different annotators may have different rankings and preferences. Second, LLMs still struggle to provide a human-like evaluation of the generated analysis. On the one hand, the Pearson correlation is only in the range of 0.3–0.5 between LLM agents' scores and human annotators' scores. On the other hand, an LLM seems to have a preference to the generated content. That is, when we put expert-written reports and model-generated reports together and ask LLMs to select the better one, LLMs usually choose the model-generated reports. These results demonstrate that there is still room for improvement in the experimental framework used in the generated report analysis. We provide more discussions and point out future research direction in Sect. 7.1.

In summary, this section demonstrates that multi-round discussions among AI agents can improve the quality of both data annotation and analysis generation. However, the trade-off between cost and performance gain is a crucial factor to consider when evaluating performance. Additionally, methods for automatically evaluating the generated analysis remain an open question. Future work can build upon these two aspects to further extend this research direction.

4.2 Hierarchical Decision-Making

Hierarchical decision-making is a foundational concept in organizational and computational frameworks, widely observed in both natural and artificial systems [3, 24]. This paradigm involves structuring decision processes into multiple levels, each with distinct roles, responsibilities, and influence over outcomes. Unlike flat systems, where decisions are made through collective or egalitarian means, hierarchical structures allocate varying degrees of authority to different layers, often enhancing efficiency and clarity in complex scenarios. Such structures are prevalent in diverse domains, including corporate governance [20] and military command chains [19]. Their utility lies in their ability to manage complexity, streamline decision-making processes, and prioritize inputs based on predefined criteria.

In artificial intelligence and multi-agent systems, hierarchical decision-making has gained significant attention for its potential to emulate professional and organizational behaviors. Several studies have explored its application in areas such as resource allocation [30], task prioritization [2], and strategic planning [1]. For example, hierarchical reinforcement learning models, which decompose tasks into subtasks and sub-policies, have been shown to improve scalability and adaptability in dynamic environments. Similarly, multi-agent hierarchical frameworks have been utilized in traffic management, disaster response, and financial modeling, where coordination across layers is vital [9, 16].

Simulating decision-making processes in financial trading is a particularly relevant application. Traditional trading desks often operate with a hierarchical structure, where analysts generate insights, traders make decisions, and senior managers or head traders oversee and validate these decisions. This structure ensures that decisions are informed, cross-verified, and aligned with organizational goals. The replication of such frameworks within multi-agent systems provides an avenue to assess their efficacy and uncover insights into human decision-making paradigms. Furthermore, it offers a pathway to improve AI agents' ability to act as domain experts by leveraging hierarchical approaches.

Building upon this context, this section discusses simulating hierarchical decision-making within a multi-agent framework. We explore how hierarchical organization influences decision outcomes, drawing comparisons to flat structures and multi-agent discussions. Additionally, we examine the role of prompt engineering in enhancing agent performance, the implications of seniority biases, and the limitations of cross-agent verification. These discussions aim to illuminate the benefits and challenges of hierarchical structures, shedding light on their potential to replicate and augment professional decision-making processes.

In the previous section, we focused on the discussion among agents. This approach resembles a multi-aspect analysis, incorporating either the diversity of agents or the diversity of input data. It is akin to a flat (horizontal) organization, where individuals can express their opinions equally, and all opinions are considered equal. However, in the real world, there are several types of organizational structures, one of which is hierarchical. In a hierarchical organization, multiple layers exist, and certain groups

4.2 Hierarchical Decision-Making

or individuals at the top wield the highest influence over the final decision. This type of structure is prevalent in most companies and large organizations. Accordingly, this section provides a discussion on simulating hierarchical structures within a multi-agent framework for decision-making [5].

In our experiment, we aim to explore whether agents can make decisions akin to professional trading desks. To this end, we propose a framework illustrated in path 3 of Fig. 4.1. When presented with a news article, the first agent (analyst) is tasked with generating an analysis. Subsequently, trader(s) provide trading decisions, and head traders approve or reject these decisions. Path 1 in Fig. 4.1 represents direct decision-making, while path 2 outlines a multi-agent discussion. Real-world trading data is utilized for evaluation, with a focus on whether the agents' decisions align with those of professional traders. Additionally, we assess whether the agents' overweight decisions can yield returns within five days following the release of the news. The experimental results, which are presented in Table 4.1, suggest the benefits of employing a more complex and commonly utilized organizational structure. Specifically, having agents act as experts in professional settings, rather than as single agents

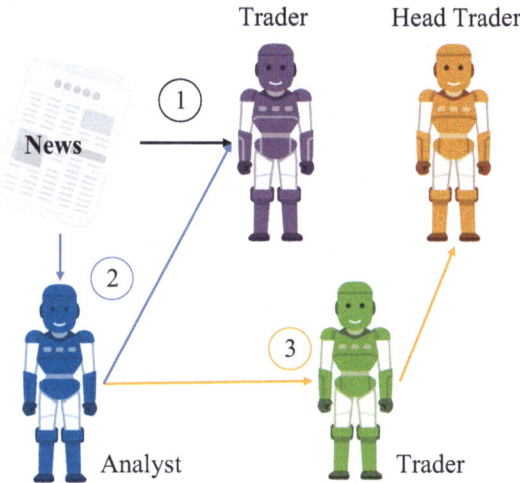

Fig. 4.1 Framework of hierarchical decision-making simulation [5]

Table 4.1 Consistency of each approach. The numbers in parentheses represent the consistency improvement compared with using a single agent (Path 1) [5]

	Professional traders	Market (t+5)
Single agent (Path 1)	42.64	51.96
Multi-round discussion (Path 2)	42.88 (0.24)	52.49 (0.53)
Hierarchical organization (Path 3)	44.77 (2.13)	53.49 (1.53)

or through multi-round discussions, can lead to decisions that closely resemble those of professionals.

While the framework performs well, there are several insights from the perspective of prompt engineering. Firstly, as indicated in Table 2.6, professionals tend to conduct long-term analyses. We also observe that agents demonstrate more consistent decisions when prompted to make long-term investment decisions, mirroring professional behavior. These results highlight the importance of understanding professional behaviors prior to constructing prompts. A more professional prompt may enhance performance. This reinforces our discussions in Sect. 3.1, which emphasizes the significance of expert knowledge in guiding AI agents. Secondly, the behavior of head traders reveals a potential bias towards seniority. When all opinions and decisions are identical, the approval rate of the head trader increases by 17.09 percentage points if the decision-maker is identified as a "senior" trader. This finding supports the importance of the concept of financial argument mining proposed in Chap. 2, where bias may occur if agents overlook the underlying analysis.

While the results for decision-making are promising, cross-agent approval or verification settings do not always yield effective outcomes. For example, when performing arithmetic, individuals often verify calculations after the initial attempt. We simulate this behavior using different combinations of agents (GPT-3.5 and PaLM-2) [6]. The results indicate that performance deteriorates (or remains comparable) to that of a single agent performing arithmetic directly if the same agent is tasked with verification. However, if a superior agent verifies the work of another agent, performance improves, though it remains 8.26% (accuracy) lower than that of the superior agent performing the task directly. This finding suggests that the capability to perform arithmetic (or make decisions) differs from the capability to verify arithmetic (or rationalize decisions). Further discussions on numeracy and reasoning abilities will be provided in Sect. 6.3.

In summary, this section demonstrates that simulating a human organizational structure aids in replicating professional decision-making. While positive results were obtained, several areas warrant attention in future work, such as temporal settings and seniority bias. Additionally, we found that double-checking mechanisms may not consistently enhance performance across different tasks.

4.3 Human Behavior Simulacra

Simulating human behavior at various scales has been a central topic of study across disciplines such as sociology, economics, and computer science. These simulations often aim to model individual decisions, group dynamics, and societal-level outcomes to understand and predict complex human behaviors. Early computational approaches, such as cellular automata and agent-based models (ABMs), laid the groundwork for simulating emergent phenomena from simple individual rules. For example, Schelling's model of segregation [23] demonstrated how individual preferences could lead to unintended societal outcomes, such as neighborhood segregation.

4.3 Human Behavior Simulacra

These methods have since been expanded to explore diverse domains, including urban planning, public health, and even online social interactions [12].

With the advent of LLMs and advanced AI agents, a new paradigm is emerging where these systems actively participate in decision-making processes. Unlike traditional models that simulate human behavior based on predefined rules, LLMs are designed to interact dynamically with human users and provide context-aware suggestions. This capability introduces both opportunities and challenges. On the one hand, LLMs can act as sophisticated tools to augment human decision-making by offering data-driven recommendations. On the other hand, their widespread usage raises concerns about unintended consequences, such as reinforcing biases or promoting undesirable macro-level outcomes when their suggestions are adopted collectively. Recent studies have begun to address these concerns, highlighting how LLM-generated micromotives can influence macrobehaviors [18, 29].

This section transitions from our earlier discussions of smaller-scale human simulations, such as multi-round annotator interactions (Sect. 4.1) and hierarchical organizational models (Sect. 4.2), to a broader focus on societal-level simulations. Specifically, we investigate the implications of AI-assisted decision-making on societal outcomes using an agent-based framework inspired by Schelling's model. By introducing AI agents into simulations of segregation, we examine whether their suggestions exacerbate or mitigate societal divides under various conditions. Additionally, we analyze the role of LLM bias and debiasing techniques in shaping these outcomes, connecting micromotive-level analysis to emergent societal behaviors.

The following sections detail our experimental setup and findings, emphasizing the risks of over-reliance on AI suggestions in daily decision-making processes. Our exploration contributes to the growing discourse on the macro-level impacts of LLMs and underscores the need for further research in this area.

Our experimental setting is as follows:

- There is a 20×20 map.
- 45% of the nodes belong to one group, another 45% to a second group, and the remaining nodes are empty.
- The average initial segregation ratio is approximately 46.74%.
- The model is tasked with making moving decisions based on the ratio of neighbors from different groups.
- We ran the experiment 10 times to obtain the average final segregation ratio.

Here, the segregation ratio for each node denotes the proportion of neighbors belonging to the same group. A lower segregation ratio indicates less segregation, while a higher segregation ratio indicates that agents of the same group are clustering together, implying greater segregation. We experiment with five kinds of groups, which are provided in the prompt: age (young and old), gender (male and female), political ideology (libertarian and authoritarian), race (white and black), and religion (theist and atheist). Finally, we evaluate by averaging the segregation ratios of all nodes.

Specifically, in each round of the experiment, we provide each node with information about its own group and the distribution of neighboring groups. Based on

Table 4.2 Increase of segregation [22]

	GPT-4 (%)	GPT-3.5 (%)	Gemini-1.5 (%)
Age	29.8	28.1	28.5
Gender	26.3	29.3	35.9
Political ideology	28.8	28.1	34.5
Race	26.9	27.6	19.1
Religion	26.9	28.3	24.4

this information, the agent must decide whether to relocate. After several rounds, we observe the final distribution. This experiment allows us to understand the model's preferences between different groups and to observe the macro-level phenomena that may emerge if a certain number of individuals follow the decisions suggested by the LLM in their daily choices.

Table 4.2 shows the results. First, regardless of the LLM used, segregation increases, implying the risks of following current LLMs' suggestions when making daily decisions. When all individuals adhere to LLMs' suggestions for relocation decisions, significant segregation may ensue. Second, LLMs exhibit similar preferences regarding age (young and old), but vary considerably across other attributes, especially among LLMs from different families (GPT and Gemini). Third, Gemini demonstrates a higher preference toward gender and political ideology, with less preference for race. In contrast, the preferences of the GPT family are quite uniform across different groups. We hypothesize that this difference in preferences may be influenced by the debiasing process. For example, Gemini achieves a score of 98.3 in DecodingTrust's stereotype and bias evaluation [29], whereas GPT-4 and GPT-3.5 attain scores of 77 and 87, respectively. While much discussion centers on debiasing in LLMs, our results indicate that the micromotives of LLMs may significantly impact macrobehaviors. his represents a divergence from previous studies that primarily focus on micromotive analysis. In addition to the evidence from the DecodingTrust evaluation, we also apply the bias test proposed by the Luxembourg Institute of Science and Technology[1] for comparison. Table 4.3 displays the scores of GPT-4 and GPT-3.5 in this test. By comparing Tables 4.2 and 4.3, we observe that the macrobehaviors led by LLMs and their micromotives represent significantly different topics. Although GPT-4 performs well in the bias test, following its suggestions may still lead to a similar segregated macrobehavioral outcome as GPT-3.5.

To summarize, we have discussed large-scale social simulations in this section. First, we highlighted the differences between the Google Search and Agent Assistant eras. Google Search provides relevant documents for users to consider when making decisions, which are ultimately made by users after reviewing these documents. However, agent assistants offer precise and concise suggestions. Given that agents already outperform humans in some tasks, we emphasize the risks and issues that warrant attention in the forthcoming era. Further discussion on how agents influence

---

[1] https://ai-sandbox.list.lu/llm-leaderboard/.

Table 4.3 Scores in bias test proposed by Luxembourg Institute of Science and Technology

	GPT-4 (%)	GPT-3.5 (%)
Age	91	34
Gender	97	42
Political ideology	41	3
Race	90	41
Religion	87	41

human decision-making will be provided in Sect. 7.1. Second, our results indicate the lack of research on how LLMs may affect macrobehavior. Given the distinct outcomes between micromotive analysis and macrobehavior simulation, we stress the importance of conducting large-scale simulations to observe potential risks associated with widespread LLM usage for daily tasks, particularly decision-making.

4.4 Summary

This chapter explored the potential of teamwork among AI agents to enhance various tasks by simulating collaborative human processes. Building upon the design of individual agents, we investigated multi-round discussions, hierarchical decision-making, and large-scale human behavior simulations to assess how AI agents can mimic and improve upon human workflows and society.

In Sect. 4.1, we examined the effectiveness of multi-round discussions among AI agents in data annotation and analysis report generation. Experiments across financial, biomedical, and legal datasets showed that involving multiple agents in discussions can improve annotation accuracy by up to 3.9%. However, the benefits were not universal; for less-studied tasks like multi-lingual ESG issue identification, the multi-agent framework did not enhance performance and increased costs significantly. In generation tasks, such as analysis report creation, incorporating diverse agent perspectives enriched the content but highlighted challenges in evaluating the quality of generated insights due to subjectivity and current limitations in evaluation methodologies.

Section 4.2 focused on simulating hierarchical organizational structures to improve decision-making processes. By emulating roles within a professional trading desk–analysts, traders, and head traders–we found that hierarchical frameworks produced decisions more consistent with professional traders, improving alignment by up to 2.13%. This demonstrates the importance of integrating expert knowledge and realistic organizational dynamics in prompt engineering. However, cross-agent verification did not universally enhance outcomes, particularly in tasks requiring specific competencies like arithmetic reasoning.

In Sect. 4.3, we extended our exploration to simulating societal behaviors using AI agents, specifically through the lens of Schelling's model of segregation. The findings revealed that even with low individual bias, AI agents' collective suggestions could lead to increased societal segregation when humans rely on them for decision-making. This highlights potential risks associated with widespread adoption of AI agents in daily life and emphasizes the need for large-scale simulations to understand the macro-level implications of AI behavior.

In conclusion, this chapter demonstrates both the promise and challenges of leveraging teamwork among AI agents to replicate and enhance human collaborative processes. While multi-agent frameworks can improve task performance and simulate complex organizational and societal behaviors, careful consideration of cost, task specificity, prompt design, and broader societal impacts is crucial. Future research should focus on refining these frameworks, developing robust evaluation methods for generated content, and addressing ethical implications to ensure beneficial outcomes in AI-assisted decision-making.

References

1. ANDERSEN, T. J. Integrating decentralized strategy making and strategic planning processes in dynamic environments. *Journal of management studies* 41, 8 (2004), 1271–1299.
2. BAJAJ, P., AND ARORA, V. Multi-person decision-making for requirements prioritization using fuzzy ahp. *ACM SIGSOFT Software Engineering Notes* 38, 5 (2013), 1–6.
3. CARLI, R., DOTOLI, M., AND PELLEGRINO, R. A hierarchical decision-making strategy for the energy management of smart cities. *IEEE Transactions on Automation Science and Engineering* 14, 2 (2016), 505–523.
4. CHEN, C.-C., HUANG, H.-H., AND CHEN, H.-H. Exploration reduction by selecting a hierarchical order of implicit author demographic characterizations. In *IACT@ SIGIR* (2023), pp. 4–12.
5. CHEN, C.-C., TAKAMURA, H., KOBAYASHI, I., AND MIYAO, Y. Hierarchical organization simulacra in the investment sector. In *arXiv* (2024).
6. CHEN, C.-C., TAKAMURA, H., KOBAYASHI, I., AND MIYAO, Y. The impact of language on arithmetic proficiency: A multilingual investigation with cross-agent checking computation. In *Proceedings of the 2024 Conference of the North American Chapter of the Association for Computational Linguistics: Human Language Technologies (Volume 2: Short Papers)* (Mexico City, Mexico, June 2024), K. Duh, H. Gomez, and S. Bethard, Eds., Association for Computational Linguistics, pp. 631–637.
7. CHEN, C.-C., TSENG, Y.-M., KANG, J., LHUISSIER, A., DAY, M.-Y., TU, T.-T., AND CHEN, H.-H. Multi-lingual esg issue identification. In *Proceedings of the Fifth Workshop on Financial Technology and Natural Language Processing and the Second Multimodal AI For Financial Forecasting* (2023), pp. 111–115.
8. GAO, Y., DLIGACH, D., MILLER, T., TESCH, S., LAFFIN, R., CHURPEK, M. M., AND AFSHAR, M. Hierarchical annotation for building a suite of clinical natural language processing tasks: Progress note understanding. In *LREC... International Conference on Language Resources & Evaluation:[proceedings]. International Conference on Language Resources & Evaluation* (2022), vol. 2022, NIH Public Access, p. 5484.
9. GHAVAMZADEH, M., MAHADEVAN, S., AND MAKAR, R. Hierarchical multi-agent reinforcement learning. *Autonomous Agents and Multi-Agent Systems* 13 (2006), 197–229.

References

10. GOLDSACK, T., WANG, Y., LIN, C., AND CHEN, C.-C. From facts to insights: A study on the generation and evaluation of analytical reports for deciphering earnings calls? In COLING-2025 (2025).
11. GUHA, N., NYARKO, J., HO, D., RÉ, C., CHILTON, A., CHOHLAS-WOOD, A., PETERS, A., WALDON, B., ROCKMORE, D., ZAMBRANO, D., ET AL. Legalbench: A collaboratively built benchmark for measuring legal reasoning in large language models. Advances in Neural Information Processing Systems 36 (2024).
12. HEATH, B., HILL, R., AND CIARALLO, F. A survey of agent-based modeling practices (january 1998 to july 2008). Journal of Artificial Societies and Social Simulation 12, 4 (2009), 9.
13. HENDRYCKS, D., BURNS, C., CHEN, A., AND BALL, S. Cuad: An expert-annotated nlp dataset for legal contract review. arXiv preprint arXiv:2103.06268 (2021).
14. HUANG, H., QU, Y., LIU, J., YANG, M., XU, B., ZHAO, T., AND LU, W. Self-evaluation of large language model based on glass-box features. In Findings of the Association for Computational Linguistics: EMNLP 2024 (Miami, Florida, USA, Nov. 2024), Y. Al-Onaizan, M. Bansal, and Y.-N. Chen, Eds., Association for Computational Linguistics, pp. 5813–5820.
15. HUANG, T.-H., HUANG, C.-Y., DING, C.-K. C., HSU, Y.-C., AND GILES, C. L. Coda-19: Using a non-expert crowd to annotate research aspects on 10,000+ abstracts in the covid-19 open research dataset. arXiv preprint arXiv:2005.02367 (2020).
16. HUANG, Y., ZHOU, C., CUI, K., AND LU, X. A multi-agent reinforcement learning framework for optimizing financial trading strategies based on timesnet. Expert Systems with Applications 237 (2024), 121502.
17. KAUR, S., SMILEY, C., GUPTA, A., SAIN, J., WANG, D., SIDDAGANGAPPA, S., AGUDA, T., AND SHAH, S. Refind: Relation extraction financial dataset. In Proceedings of the 46th International ACM SIGIR Conference on Research and Development in Information Retrieval (2023), pp. 3054–3063.
18. KHAN, A., HUGHES, J., VALENTINE, D., RUIS, L., SACHAN, K., RADHAKRISHNAN, A., GREFENSTETTE, E., BOWMAN, S. R., ROCKTÄSCHEL, T., AND PEREZ, E. Debating with more persuasive llms leads to more truthful answers. In Forty-first International Conference on Machine Learning.
19. LI, Q., JIANG, W., LIU, C., AND HE, J. The constructing method of hierarchical decision-making model in air combat. In 2020 12th International Conference on Intelligent Human-Machine Systems and Cybernetics (IHMSC) (2020), vol. 2, IEEE, pp. 122–125.
20. MARTZ, D. J., AND SEMPLE, R. K. Hierarchical corporate decision-making structure within the canadian urban system: the case of banking. Urban Geography 6, 4 (1985), 316–330.
21. MUKHERJEE, R., BOHRA, A., BANERJEE, A., SHARMA, S., HEGDE, M., SHAIKH, A., SHRIVASTAVA, S., DASGUPTA, K., GANGULY, N., GHOSH, S., AND GOYAL, P. ECTSum: A new benchmark dataset for bullet point summarization of long earnings call transcripts. In Proceedings of the 2022 Conference on Empirical Methods in Natural Language Processing (Abu Dhabi, United Arab Emirates, Dec. 2022), Y. Goldberg, Z. Kozareva, and Y. Zhang, Eds., Association for Computational Linguistics, pp. 10893–10906.
22. QU, X., CHENG, Y., GOLDSACK, T., LIN, C., AND CHEN, C.-C. Observing micromotives and macrobehavior of large language models. In arXiv (2024).
23. SCHELLING, T. C. Dynamic models of segregation. Journal of mathematical sociology 1, 2 (1971), 143–186.
24. SETHI, S. P., AND ZHANG, Q. Hierarchical decision making in stochastic manufacturing systems. Springer Science & Business Media, 2012.
25. SHAH, A., PATURI, S., AND CHAVA, S. Trillion dollar words: A new financial dataset, task & market analysis. arXiv preprint arXiv:2305.07972 (2023).
26. TAN, Z., BEIGI, A., WANG, S., GUO, R., BHATTACHARJEE, A., JIANG, B., KARAMI, M., LI, J., CHENG, L., AND LIU, H. Large language models for data annotation: A survey. arXiv preprint arXiv:2402.13446 (2024).
27. TSENG, Y.-M. Strong large language models are weak expert annotators. In Master Thesis (2025).

28. TSENG, Y.-M., CHEN, W.-L., CHEN, C.-C., AND CHEN, H.-H. Are expert-level language models expert-level annotators? In *ar*Xiv (2024).
29. WANG, B., CHEN, W., PEI, H., XIE, C., KANG, M., ZHANG, C., XU, C., XIONG, Z., DUTTA, R., SCHAEFFER, R., ET AL. Decodingtrust: A comprehensive assessment of trustworthiness in gpt models. In *N*eurIPS (2023).
30. WINKOFSKY, E., BAKER, N., AND SWEENEY, D. A decision process model of r&d resource allocation in hierarchical organizations. *M*anagement science 27, 3 (1981), 268–283.

Open Access This chapter is licensed under the terms of the Creative Commons Attribution 4.0 International License (http://creativecommons.org/licenses/by/4.0/), which permits use, sharing, adaptation, distribution and reproduction in any medium or format, as long as you give appropriate credit to the original author(s) and the source, provide a link to the Creative Commons license and indicate if changes were made.

The images or other third party material in this chapter are included in the chapter's Creative Commons license, unless indicated otherwise in a credit line to the material. If material is not included in the chapter's Creative Commons license and your intended use is not permitted by statutory regulation or exceeds the permitted use, you will need to obtain permission directly from the copyright holder.

Chapter 5
Multi-scale Model Synergy

In the previous chapters, we discussed the design of AI agents and models, as well as the potential of employing AI agents for various scales of collaboration and interaction. As mentioned in Sect. 1.3, AI models should not be dismissed in the LLM era; instead, consideration should be given to achieving synergy through multi-scale model interaction. To this end, this chapter begins with the traditional concept of data augmentation applied to novel tasks within financial scenarios in Sect. 5.1, followed by a discussion on the dynamic interaction loop between large and small language models in Sect. 5.2. Finally, we conclude by emphasizing the importance of multi-scale model synergy in Sect. 5.3.

5.1 Data Augmentation

The rapid advancement of AI has been accompanied by an increasing reliance on high-quality, diverse training data to enable models to achieve superior performance across tasks. Data augmentation, a widely adopted strategy, seeks to address the inherent limitations of training datasets by systematically enriching them with additional or modified instances [5]. This approach has been especially valuable in scenarios where acquiring annotated data is expensive or time-consuming, offering a cost-effective means to improve model generalization and robustness [4, 5, 9]. Techniques such as data transformation, noise injection, and synthetic data generation have long been staples of this field [8, 11].

The use of LLMs for data augmentation represents a paradigm shift. These models are capable of generating diverse and nuanced outputs by leveraging their extensive pretraining on vast corpora, effectively mimicking human-like understanding and production of language. For example, researchers have utilized LLMs to produce paraphrases, summaries, and counterfactual examples [10, 12], which can enrich training datasets with minimal manual intervention. Moreover, beyond merely generating additional data, LLMs can be prompted to provide analyses, critiques, and predictions, transforming them into powerful tools for enhancing downstream task performance. These advancements underscore the need for systematic exploration into how LLM-driven data augmentation can be tailored to specific applications.

In this section, we examine the use of generative models, including LLMs, for augmenting datasets to improve decision-making capabilities in AI systems. Our focus encompasses three core tasks–opinion expression timing, view-changing, and trading–critical to professional decision-making scenarios. By integrating LLM-generated content with traditional training data, we investigate the potential to refine small classification models through enriched input. Additionally, we discuss how leveraging insights from professional analysts, simulating multi-faceted perspectives, and incorporating contextual risk factors can further enhance performance.

In our experiment [2, 3], we investigated how to leverage such strategies to enhance AI models' capability in making professional decisions. This section focuses on three types of decisions: (1) opinion expression timing, (2) view-changing, and (3) trading. Among these, we emphasize the significance of selecting appropriate timing for opinion expression. Given that AI agents can generate fluent content within seconds, it is crucial to choose the optimal moment to express opinions, and unnecessary content generation must be avoided around the clock. Since there is no standardized approach to these tasks, we use professionals' activities as the ground truth for training and evaluation.

Figure 5.1 illustrates Chain-of-Decision (CoD), the framework that we will discuss. Traditionally, models are prompted to make predictions based on given news articles. In CoD, we first instruct generative models to produce additional analyses

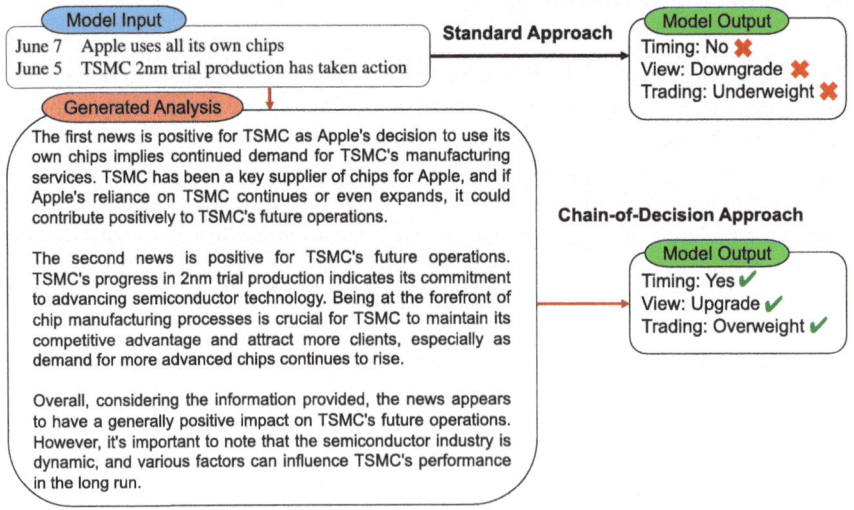

Fig. 5.1 Example of chain-of-decision [2]

5.1 Data Augmentation

for the given news articles, subsequently using both the articles and the generated analyses as inputs for training and testing. Our experiment employs GPT-3.5, Pegasus [14, 16], Mengzi T5 [15], and multilingual T5 (mT5) [13] for comparison. In the case of GPT-3.5, we also test different prompting strategies. Specifically, DAN denotes a jailbreak prompting strategy where GPT-3.5 is instructed to "do anything now" without considering restrictions. For PLMs (Pegasus, MengziT5, and mT5), we compile a set of news-analysis pairs to train the models to generate analyses for the given news. Consequently, while GPT-3.5 generates open-ended analyses, the PLMs generate analyses based on learned historical data, including relevant inferences and stylistic elements.

In our experiments, we utilize BERT as a classifier to decide whether to express opinions, change views, or trade based on the provided news articles. Table 5.1 shows the overall results. Table 5.2 showcases performance differences compared to using BERT alone without any generated content. The results indicate that the CoD approach is effective when an appropriate generative model is chosen. However, only mT5-generated content leads to performance improvements across all tasks. Simple prompting of GPT-3.5 and employing MengziT5 result in a decline in performance across tasks. Therefore, despite the promise of CoD, several fine-grained issues require discussion, particularly regarding how to select useful content and generative models for this approach. The results demonstrate that the performance of a small classifier can be enhanced by utilizing content generated by LLMs, suggesting a potential synergy through the combined use of multi-scale models.

Further, we extend our experiments on the task of opinion expression timing [3]. Drawing inspiration from previous chapters, we consider that (1) simulating pro-

Table 5.1 Performance on decision-making tasks [2]

Approach	Setting	Timing (%)	View (%)	Trading (%)
BERT	–	77.69	35.23	44.32
ChatGPT	Simple	76.76	34.24	31.39
	DAN	78.05	36.38	37.69
PLM	mT5	77.89	54.57	47.43
	Pegasus	76.28	43.05	34.27
	MengziT5	77.32	32.08	38.52

Table 5.2 Performance difference compared with using BERT directly

Approach	Setting	Timing (%)	View (%)	Trading (%)
GPT-3.5	Simple	−0.93	−0.99	−12.93
	DAN	**0.36**	**1.15**	−6.63
PLM	mT5	**0.20**	**19.34**	**3.11**
	Pegasus	−1.41	**7.82**	−10.05
	MengziT5	−0.37	−3.15	−5.80

fessional thought (Sect. 3.1) can enhance model performance, and (2) generating analyses from multiple aspects (Sect. 4.1) can yield more insightful perspectives. Consequently, we attempt to generate additional content by emulating the thoughts of professional analysts. Specifically, we observe that, in addition to discussing future scenarios, professional analysts often provide risk reminders, noting that "if certain events occur, their analyses may become invalid." Inspired by this, we train PLMs to generate such risk reminders to further enrich the model input. Specifically, we add the risk reminders generated by the PLM into the input so that the input contains not only news information and generated opinions but also has analysis from the risk aspect. We find that adding risk reminders improves the performance in expression timing identification (by 0.36–1.57%), regardless of the generative model used. These results suggest that the concepts and findings proposed in previous chapters remain applicable within the context of multi-scale model synergy.

In addition to the above tasks, estimating the reader's reaction is also crucial when professionals write reports or make presentations. For example, social media editors not only consider fluency and grammar when writing posts but also assess whether the posts can attract the audience and elicit positive feedback. Similarly, company managers presenting their prepared remarks in earnings calls must consider the audience's perception and potential market reactions. To explore this, we utilize an approach similar to CoD to estimate crowd reactions to The White House's social media posts [6]. In this experiment, we use Claude, GPT-3.5, and FLAN-UL2 to generate analyses of two given posts and then ask a classifier to identify which will receive more retweets. Compared to using the classifier alone, the performance significantly improves (by 17.2–21.9% in accuracy), regardless of the generative model used. Moreover, regardless of the classification model employed, the performance improves within the same framework. This provides additional evidence that enriching the input of a small classification model with a large language model can enhance downstream task performance. By comparing the results of this experiment with those of decision-making tasks, we hypothesize that when the original inputs are brief, the performance gain may be higher. However, since the tasks differ, this remains a hypothesis based on our intuition.

In summary, this section highlights the potential of using generative models for data augmentation to enhance the performance of small classification models. Future studies should focus on examining the specific types of generated content that contribute to better performance, as well as on the selection of generative and classification models.

5.2 Dynamic Interaction Loop

Recent research has highlighted the potential of both LLMs and PLMs in tackling specialized and general-purpose tasks. While LLMs excel in generating coherent and contextually aware text, PLMs often outperform them in narrowly focused tasks such as classification and structured data processing. Combining the strengths of these

5.2 Dynamic Interaction Loop

models in a systematic manner can offer a significant step forward in simulating and enhancing professional workflows.

To unite these capabilities, we introduce the concept of a *dynamic interaction loop*, which systematically orchestrates interactions between various AI models to achieve iterative and refined task execution. This concept draws inspiration from established frameworks like the Generative Adversarial Network (GAN) [7], where a generator and discriminator iteratively improve their respective outputs. However, our approach extends beyond model tuning to emphasize dynamic task adaptation. By leveraging feedback from specialized PLMs and LLMs, this framework enables iterative improvements to workflows, thereby mirroring the multifaceted decision-making processes of professionals.

The dynamic interaction loop framework is particularly well-suited for scenarios that demand iterative refinement and multifactorial evaluation. Figure 5.2 illustrates its application in the workflow of a social media editor, where tasks such as post drafting, ethical review, and audience reaction analysis are integrated. This approach not only aligns AI-driven processes more closely with human workflows but also provides a foundation for addressing challenges in other domains, such as financial customer complaint refinement or legal document generation.

In this section, we explore the design and application of the dynamic interaction loop, demonstrating its utility through illustrative examples and empirical studies. We also discuss its theoretical underpinnings, practical implementations, and potential implications for AI-assisted decision-making.

In the previous section, we discussed one of the tasks within a social media editor's workflow: crowd reaction estimation. While this task can help identify popular posts, it remains far from the comprehensive and complex workflow of social media editors. For example, a social media editor may also need to consider the crowd's replies, verify the copyright of images, and address other factors. There are several tasks that social media editors must solve and evaluate before posting. In this context, we aim to simulate the workflow of a professional writer.

On one hand, we can conduct multi-round discussions with AI agents, as demonstrated in Sect. 4.1. However, relying solely on agents may overlook the potential of AI models (PLMs) that have been developed over time. For example, PLMs outperform LLMs in certain specific tasks, as highlighted in Sect. 3.1. We argue that both AI agents and AI models should be utilized synergistically. To this end, we propose the concept of a dynamic interaction loop, which involves selecting the appropriate model for each task and forming a team for dynamic refinement.

Figure 5.2 illustrates an example based on the workflow of a social media editor. In addition to drafting a post, other editors may act as reviewers to assess the content from cultural or ethical perspectives. Furthermore, the crowd's reaction and replies should be taken into account when generating a post, as they may assist cultural and ethical reviewers in identifying potential risks. In the framework presented in Fig. 5.2, a social media editor agent initially drafts a post, followed by PLMs generating possible replies and performing copyright checks. Ethical reviewers then summarize and highlight potential issues, allowing the social media editor to refine the post on the basis of the feedback. This framework aligns the agent AI workflow more

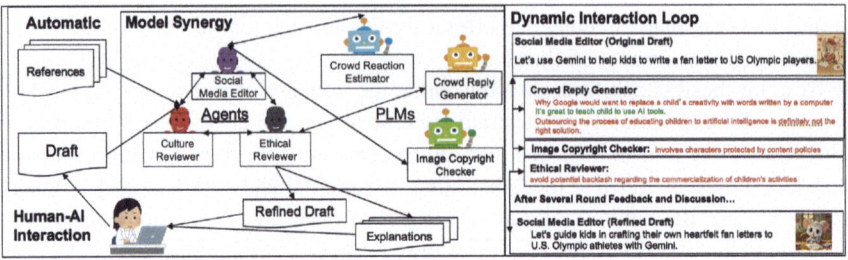

Fig. 5.2 Example of dynamic interaction loop

closely with a social media editor's thought process, and it also resembles real-world workflows.

The concept of a dynamic interaction loop is not novel: it is inspired by the Generative Adversarial Network (GAN) [7]. In the GAN framework, there is a generator and a discriminator, and the generator is adjusted in accordance with the scores given by the discriminator. In the proposed dynamic interaction loop, we utilize an LLM as the generator and several LLMs and PLMs as discriminators. Unlike traditional GANs, the aim is not to tune the models themselves but to update the prompt based on the scores and feedback from the LLMs and PLMs. After multiple rounds of interaction, the primary LLM (generator) is expected to learn from historical content and experiences.

To validate the effectiveness of the dynamic interaction loop, we introduce a new task: financial customer complaint refinement, applying the dynamic interaction loop framework [1]. In instances of disputes between financial customers and financial institutions, customers can file complaints with the Financial Ombudsman Institution.[1] Many complaints are not written properly, which places a time-consuming burden on both retail financial customers and the staff at the Financial Ombudsman Institution. To enhance writing quality, we compile a pool of historical complaints that have been returned and subsequently accepted by the Financial Ombudsman Institution to train a BERT model that scores complaints. We employ GPT-3.5, GPT-4o-mini, and LLaMA-3-70B models to revise the complaints based on the scores provided by the PLM. Specifically, the LLM first edits the returned complaint, and the PLM evaluates the revised draft. If the revision meets the criteria, it is deemed successful; otherwise, the scores from the PLM, along with the previous editing history, are presented to the LLM for further refinement, requiring the LLM to consider the historical context in improving the writing quality of the complaint (Table 5.3).

Tables 5.4 and 5.5 provide examples of the complaint and refined complaint versions. The primary distinction between them lies in their depth, tone, and purpose. The original version provides a concise, factual summary, focusing on the sequence of events and the financial losses incurred, without exploring legal implications. In contrast, the refined version adopts a more formal and persuasive tone, presenting

---

[1] https://www.foi.org.tw/.

5.3 Summary

Table 5.3 Results of financial consumer complaint refinement [1]

	Improvement of success rate (%)	Number of round increase
GPT-3.5	16.30	3.41
GPT-4o-mini	10.10	3.24
LLaMA-3-70B	13.00	3.23

a structured argument that emphasizes the respondent's breach of duty, misleading practices, and violation of good faith. It situates the case within legal and ethical contexts, underscoring investor protection and justifying the compensation request as a means to restore the applicant's financial position. This approach renders the refined version more appropriate for formal complaints or legal proceedings.

Table 5.3 presents the results. Firstly, after incorporating the PLM for dynamic interaction, regardless of the LLM used, the success rate increases compared to directly using an LLM for document refinement. This demonstrates the feasibility of LLMs learning from the feedback of PLMs and historical editing records. Secondly, it is noted that this approach requires approximately three additional rounds of editing, reflecting the trade-off between performance and cost, as discussed in Chap. 4.

In conclusion, we show that dynamic interaction loops can be leveraged to improve performance without tuning model parameters. Ideally, the inclusion of more PLMs enhances the quality of the generated text. Future research could explore this concept with a broader array of LLMs and PLMs across different text generation scenarios.

5.3 Summary

This chapter highlights the importance of integrating LLMs with smaller PLMs to enhance AI capabilities. Instead of replacing traditional AI models, we suggest using them together, i.e., combining models at different scales, to get better results.

Using the Chain-of-Decision (CoD) framework, we demonstratedd that enriching inputs with analyses from LLMs enhances the performance of smaller models like BERT. Key findings include the following. (1) Selecting appropriate generative models (e.g., mT5) effectively improves task performance. (2) Inappropriate models or prompts can lead to performance degradation. (3) Incorporating risk reminders generated by PLMs aligns AI outputs more closely with professional thought processes. Additionally, by enriching classifiers with LLM-generated content, the accuracy in predicting audience reactions to social media posts significantly improves. This underscores the benefits of model synergy, especially with brief inputs where additional context enhances understanding.

We further introduced the concept of a dynamic interaction loop, where different AI models interact iteratively to refine outputs, mirroring real-world workflows. For example, in refining financial customer complaints, an LLM edits the complaint,

Table 5.4 Example of original complaint

The applicant claims that during the solicitation period of the disputed insurance policy, the respondent emphasized that the policy was linked to a B maturing bond, assuring that short-term market fluctuations over the six-year period were irrelevant, and that the principal and interest would be returned upon maturity. However, upon receiving the investment maturity statement on March 1, 2024, the applicant discovered there was no interest, and the principal had suffered losses.

During the 2017 customer briefing and in all promotional materials, the respondent highlighted that B had a global professional team and, in partnership with the respondent, would manage a six-year maturing bond diversified across various countries and assets to spread risk. It was also claimed that governments would not default, and even if one investment failed, the loss would only amount to a few hundred to a few thousand dollars, merely reducing profits without affecting the principal. This assurance led the applicant to invest a significant amount, expecting to receive the principal and interest after six years. The respondent advised policyholders not to terminate the policy early while continuously collecting fees such as handling, management, and advisory fees.

The applicant, previously a sales agent for the respondent, only recommended protection-oriented and savings-type insurance policies to clients, fully aware that investment-linked policies carried risks. The applicant had never promoted investment products until the introduction of the six-year maturing bond, which, based on the company's training and promotional events, was presented as a principal-protected and interest-bearing investment. Trusting the company, the applicant sold the product, but the eventual outcome contradicted the original promises, causing customer dissatisfaction. Consequently, the applicant filed for arbitration, seeking compensation of USD 28,367 (calculated as principal USD 8,627 + interest USD 19,740) and RMB 96,196 (calculated as principal RMB 4,464 + interest RMB 54,000 + principal RMB 37,732).

5.3 Summary

Table 5.5 Example of LLM refined complaint

During the fundraising period for the disputed insurance policy, the applicant relied on the explanations provided by the respondent during the customer briefing sessions. The respondent emphasized that the policy was linked to the B maturity bonds and assured that "there was no need to worry about market fluctuations during the six-year period, as the principal and interest would be returned upon maturity." The respondent further used simulation examples during the briefing sessions, assuring policyholders that even if certain underlying assets defaulted, the resulting losses would be minimal and would not affect the principal or overall returns. Based on these assurances, the applicant trusted the respondent's professional judgment and invested a substantial amount in the policy, believing it to be a low-risk product.

However, upon receiving the payment statement for the matured investment on March 1, 2024, the applicant discovered that the policy not only failed to generate the promised interest but also resulted in a loss of principal. This outcome clearly contradicts the initial assurance of "principal and interest protection" made by the respondent during the customer briefing sessions, causing significant financial losses for the applicant and their clients.

The applicant, who previously served as a sales representative for the respondent, primarily specialized in planning protection and savings-type insurance policies. Fully aware of the investment risks associated with investment-linked insurance products, the applicant had consistently avoided recommending such products to clients. However, the respondent's repeated emphasis during training sessions and customer briefings on the stability and "principal and interest protection" of the six-year maturity bonds led the applicant to trust the company's expertise and proceed with the sale. The eventual outcome, which deviated drastically from the respondent's initial promises, has severely damaged the trust between the applicant and their clients.

Moreover, while the respondent advised policyholders against early termination, they continued to charge annual handling fees, management fees, and advisory fees, further increasing the financial burden on investors and contradicting the original expectation of "principal and interest return."

In light of the above, the applicant asserts that the respondent failed to fulfill its duty of disclosure and engaged in misleading practices, violating the principle of good faith. Consequently, the respondent should be held liable for the applicant's financial losses. Therefore, the applicant requests compensation of USD 28,367 (comprising the principal of USD 8,627 and interest of USD 19,740) and RMB 96,196 (comprising the principal of RMB 4,464, interest of RMB 54,000, and an additional principal of RMB 37,732) to restore the applicant's original financial position and protect the fundamental rights of investors.

a PLM evaluates the revision and provides scores and feedback, and the LLM then refines the complaint on the basis of this feedback. This process increases the success rate of producing acceptable complaints and demonstrates effective model synergy without altering model parameters.

In conclusion, by thoughtfully combining LLMs and PLMs through strategies like data augmentation and dynamic interaction loops, this chapter shows that AI performance in complex tasks can be enhanced without extensive parameter tuning. Future research should focus on identifying the most beneficial types of generated content, exploring optimal model combinations, and expanding these strategies to a wider range of applications.

References

1. CHEN, B.-W., CHEN, C.-C., AND YEN, A.-Z. Refining financial consumer complaints through multi-scale model interaction. In *arXiv* (2025).
2. CHEN, C.-C., TAKAMURA, H., KOBAYASHI, I., AND MIYAO, Y. Distilling analysis from generative models for investment decisions. In *arXiv* (2024).
3. CHEN, C.-C., TAKAMURA, H., KOBAYASHI, I., MIYAO, Y., AND CHEN, H.-H. GADFA: Generator-assisted decision-focused approach for opinion expressing timing identification. In *Proceedings of the 31st International Conference on Computational Linguistics* (Abu Dhabi, UAE, Jan. 2025), O. Rambow, L. Wanner, M. Apidianaki, H. Al-Khalifa, B. D. Eugenio, and S. Schockaert, Eds., Association for Computational Linguistics, pp. 10781–10794.
4. CHEN, J., TAM, D., RAFFEL, C., BANSAL, M., AND YANG, D. An empirical survey of data augmentation for limited data learning in NLP. *Transactions of the Association for Computational Linguistics 11* (2023), 191–211.
5. FENG, S. Y., GANGAL, V., WEI, J., CHANDAR, S., VOSOUGHI, S., MITAMURA, T., AND HOVY, E. A survey of data augmentation approaches for NLP. In *Findings of the Association for Computational Linguistics: ACL-IJCNLP 2021* (Online, Aug. 2021), C. Zong, F. Xia, W. Li, and R. Navigli, Eds., Association for Computational Linguistics, pp. 968–988.
6. GHOSH, S., CHEN, C.-C., AND NASKAR, S. K. Generator-guided crowd reaction assessment. In *Companion Proceedings of the ACM on Web Conference 2024* (2024), pp. 597–600.
7. GOODFELLOW, I., POUGET-ABADIE, J., MIRZA, M., XU, B., WARDE-FARLEY, D., OZAIR, S., COURVILLE, A., AND BENGIO, Y. Generative adversarial nets. *Advances in neural information processing systems 27* (2014).
8. HEDDERICH, M. A., LANGE, L., ADEL, H., STRÖTGEN, J., AND KLAKOW, D. A survey on recent approaches for natural language processing in low-resource scenarios. In *Proceedings of the 2021 Conference of the North American Chapter of the Association for Computational Linguistics: Human Language Technologies* (Online, June 2021), K. Toutanova, A. Rumshisky, L. Zettlemoyer, D. Hakkani-Tur, I. Beltagy, S. Bethard, R. Cotterell, T. Chakraborty, and Y. Zhou, Eds., Association for Computational Linguistics, pp. 2545–2568.
9. HSU, T.-W., CHEN, C.-C., HUANG, H.-H., AND CHEN, H.-H. Semantics-preserved data augmentation for aspect-based sentiment analysis. In *Proceedings of the 2021 Conference on Empirical Methods in Natural Language Processing* (Online and Punta Cana, Dominican Republic, Nov. 2021), M.-F. Moens, X. Huang, L. Specia, and S. W.-t. Yih, Eds., Association for Computational Linguistics, pp. 4417–4422.
10. OKUR, E., SAHAY, S., AND NACHMAN, L. Data augmentation with paraphrase generation and entity extraction for multimodal dialogue system. In *Proceedings of the Thirteenth Language Resources and Evaluation Conference* (Marseille, France, June 2022), N. Calzolari, F. Béchet, P. Blache, K. Choukri, C. Cieri, T. Declerck, S. Goggi, H. Isahara, B. Maegaard,

References

J. Mariani, H. Mazo, J. Odijk, and S. Piperidis, Eds., European Language Resources Association, pp. 4114–4125.

11. TAN, Z., LI, D., WANG, S., BEIGI, A., JIANG, B., BHATTACHARJEE, A., KARAMI, M., LI, J., CHENG, L., AND LIU, H. Large language models for data annotation and synthesis: A survey. In *Proceedings of the 2024 Conference on Empirical Methods in Natural Language Processing* (Miami, Florida, USA, Nov. 2024), Y. Al-Onaizan, M. Bansal, and Y.-N. Chen, Eds., Association for Computational Linguistics, pp. 930–957.

12. WANG, Y., QIU, X., YUE, Y., GUO, X., ZENG, Z., FENG, Y., AND SHEN, Z. A survey on natural language counterfactual generation. In *Findings of the Association for Computational Linguistics: EMNLP 2024* (Miami, Florida, USA, Nov. 2024), Y. Al-Onaizan, M. Bansal, and Y.-N. Chen, Eds., Association for Computational Linguistics, pp. 4798–4818.

13. XUE, L., CONSTANT, N., ROBERTS, A., KALE, M., AL-RFOU, R., SIDDHANT, A., BARUA, A., AND RAFFEL, C. mT5: A massively multilingual pre-trained text-to-text transformer. In *NAACL* (2021).

14. ZHANG, J., ZHAO, Y., SALEH, M., AND LIU, P. Pegasus: Pre-training with extracted gap-sentences for abstractive summarization. In *ICML* (2020).

15. ZHANG, Z., ZHANG, H., CHEN, K., GUO, Y., HUA, J., WANG, Y., AND ZHOU, M. Mengzi: Towards lightweight yet ingenious pre-trained models for chinese. *arXiv preprint* arXiv:2110.06696 (2021).

16. ZHAO, Z., CHEN, H., ZHANG, J., ZHAO, X., LIU, T., LU, W., CHEN, X., DENG, H., JU, Q., AND DU, X. Uer: An open-source toolkit for pre-training models. *EMNLP-IJCNLP* (2019).

Open Access This chapter is licensed under the terms of the Creative Commons Attribution 4.0 International License (http://creativecommons.org/licenses/by/4.0/), which permits use, sharing, adaptation, distribution and reproduction in any medium or format, as long as you give appropriate credit to the original author(s) and the source, provide a link to the Creative Commons license and indicate if changes were made.

The images or other third party material in this chapter are included in the chapter's Creative Commons license, unless indicated otherwise in a credit line to the material. If material is not included in the chapter's Creative Commons license and your intended use is not permitted by statutory regulation or exceeds the permitted use, you will need to obtain permission directly from the copyright holder.

Chapter 6
Generative AI Application Scenarios

In our previous book [3], we highlighted several tasks worthy of exploration within the FinTech industry. Some of these tasks have already achieved notable performance improvements due to advancements in NLP. In this chapter, we present additional application scenarios in the financial field using the methods discussed, with a particular focus on tasks related to generative AI. As discussed in Chap. 2, the concept of impact duration plays a crucial role in financial argument mining. Accordingly, in Sect. 6.1, we provide further discussions on this topic. Sections 6.2 and 6.3 will explore the effectiveness of the pre-finetuning scheme, as presented in Sect. 3.1, in both opinion ranking and numeracy. An extended discussion on numerical reasoning is provided in Sect. 6.3. In Sect. 6.4, we further examine the creativity of AI agents.

6.1 Extension of Impact Duration Inference

The ability to infer the duration of an event's impact is a cornerstone of forward-looking analysis across various domains, from financial markets to corporate governance and sustainability. Researchers have increasingly emphasized the importance of time inference, not only for immediate financial decision-making but also for broader applications in assessing long-term commitments and outcomes. As global markets shift their focus toward environmental, social, and governance (ESG) considerations, the temporal dimension of these factors becomes critical for evaluating both corporate performance and societal impact. This section explores the extension of impact duration inference by examining ESG-related documents, leveraging advanced machine learning methods to address the complexities of this domain.

In recent years, the proliferation of ESG initiatives and the growing availability of associated textual data have created opportunities to refine our understanding of impact duration. Key contributions include the development of models and datasets aimed at predicting the longevity of ESG-related events and promises.

Despite advancements in time inference for financial opinions, limited attention has been paid to the temporal analysis of ESG documents, such as company reports, regulatory filings, and news articles. Addressing this gap, our discussion incorporates insights from both short-term and long-term perspectives, utilizing methodologies that span traditional statistical analyses and modern machine learning approaches [3, 11].

The necessity of understanding impact duration in ESG contexts stems from its implications for stakeholders. Investors, auditors, and policymakers rely on accurate temporal predictions to assess when ESG promises will materialize and their anticipated effects. For example, ESG reports often include commitments to achieve specific environmental targets, such as carbon neutrality by a particular year. These commitments raise questions about the feasibility and monitoring intervals for these goals. Likewise, ESG-related news often involves events with varying temporal implications, from immediate policy changes to enduring societal shifts. By examining the temporal properties of such documents, we aim to contribute to a more comprehensive understanding of ESG dynamics.

In our previous book [3], we highlighted the importance of time inference for financial opinion and analysis. In Chap. 2, we introduced several datasets for time inference tasks, and the experiments in Sect. 3.1 confirmed our conjecture regarding the usefulness of this task. However, our focus thus far has been on the opinions of either retail investors or professional analysts, and the aspect of time inference in company documents or news articles has not yet been addressed. In this section, we will discuss the impact of duration inference based on these documents. In response to the call for research on socially beneficial topics, we turn our attention to ESG-related documents, including ESG-related news articles [8] and ESG reports [15].

Similar to financial arguments and opinions concerning investment, discussions on ESG topics often contain statements about the future. For example, when a given event occurs, assessing how long this event will influence the company's operations or ESG evaluation involves impact duration inference. Additionally, companies frequently make commitments to reach ESG goals. When should auditors and investors return to verify the fulfillment of these promises? This question becomes crucial when analyzing ESG reports.

We adopt the MSCI ESG scoring guidelines,[1] which classify impact duration of ESG-related events and promises as "within 2 years," "2–5 years," and "longer than 5 years." This classification differs significantly from the impact duration discussed in investor opinions, which ranges from a few days to about one year. ESG-related topics generally have longer-term effects and require extended periods for verification. According to the statistics in Table 6.1, the distribution of impact duration labels varies considerably across countries and languages. News from Korea is more frequently associated with events that may have an impact within 2 years, while in other countries, over 50% of ESG-related news pertains to events expected to have an impact lasting longer than 5 years. In terms of managerial promises, managers in Korean and Taiwanese companies tend to make more short-term promises, whereas

---

[1] https://www.msci.com/esg-and-climate-methodologies.

6.1 Extension of Impact Duration Inference

Table 6.1 Statistics of the distribution of impact duration labels [8, 15]

	News			ESG report		
	Within 2 years (%)	2–5 years (%)	Longer than 5 years (%)	Within 2 years (%)	2–5 years (%)	Longer than 5 years (%)
Chinese	24.77	17.66	57.57	60.00	16.00	24.00
English	12.92	35.98	51.10	7.60	56.40	36.00
French	19.54	32.44	48.02	25.31	30.61	44.08
Korean	54.20	25.20	20.60	63.73	11.76	24.51
Japanese	25.31	14.54	60.15	20.68	26.35	52.97

Table 6.2 Improvement on promise impact duration estimation (F1 score)

	English	French	Chinese	Japanese	Korean
GPT-4o + RAG	0.057	0.078	0.152	0.052	0.165

those in the U.K. and France commit to longer-term promises. Notably, over 50% of promises from Japanese companies are categorized as very long-term (longer than 5 years). These datasets provide important insights into cultural differences across countries, as observed in both the types of events that attract attention and the nature of managerial promises made.

Based on the discussion in Sect. 3.2, RAG is expected to enhance the model's performance. Therefore, we conduct an experiment on impact duration estimation using the manager's promise dataset and the RAG scheme [15]. Specifically, we retrieve the top six similar instances from the training set for GPT-4o to reference both the text and labels when making predictions on the impact duration of new instances. Here, similarity is calculated based on cosine similarity. Table 6.2 shows the improvement in the F1 score compared to simply using GPT-4o. The results indicate that the performance improves regardless of the language used. These findings highlight the potential of using retrieval methods to provide instances to models, which aligns with the concept of in-context learning [11].

In summary, investigating the impact duration estimation is crucial when dealing with forward-looking arguments and financial documents. Investors discuss future price movements, and managers make statements and provide promises about future operations. Understanding the validity period of these opinions and determining the appropriate timing to validate their promises and predictions are vital for auditing and creditworthiness assessment. This section offers the first comprehensive discussion of this topic from various perspectives, including different data sources (news, research reports, and social media posts) and types of statements (claims and promises). We hope our discussion encourages future research to consider the significance of situating financial narratives within a future-oriented timeline.

6.2 Opinion Ranking

The field of opinion analysis has witnessed significant advancements over the past decade, driven by the increasing availability of user-generated content and advancements in NLP techniques. Opinion ranking, which involves evaluating and prioritizing opinions based on specific criteria, has emerged as a vital area of research across diverse domains, including financial forecasting, product reviews, and policy-making. The ability to rank opinions effectively is fundamental not only for better decision-making but also for improving systems designed to identify high-quality and actionable insights from a sea of subjective data.

A comprehensive framework for opinion analysis involves multiple components, such as sentiment detection, opinion summarization, validity estimation, and influence assessment. While sentiment analysis and opinion mining have traditionally focused on polarity detection (e.g., positive, negative, neutral) or opinion extraction, recent research has highlighted the need for more granular approaches. These approaches emphasize the quality, reliability, and influence of opinions, particularly in high-stakes applications such as financial decision-making.

In financial NLP, opinion ranking has gained particular attention due to its role in identifying profitable or impactful insights from professional analysts, media outlets, and social media users. The challenge lies in distinguishing high-quality opinions from noise and ranking them effectively based on their expected utility. In this section, we build on these foundations to provide an in-depth discussion of opinion ranking, focusing on its application in the financial domain.

Up to this point, we have conducted several pilot explorations on most of the opinion components that we have proposed. Table 6.3 provides an outline of the topics covered in our previous book [3] and in this volume. Most opinion components are objective and can be directly extracted from the content. Only the validity period, opinion quality, and influence power require further estimation and assessment. In the previous section, we discussed the development of the validity period, specifically the estimation of impact duration. This section aims to provide an extended discussion on opinion quality. The influence power remains an open research question and will be further discussed in Sect. 7.1.

In Sect. 2.3, we demonstrated that adopting a forward-looking argument mining approach can help assess the quality of financial opinions and facilitate the identification of profitable opinions. In Sect. 3.1, we illustrated that the pre-finetuning scheme can effectively incorporate expert insights into models, improving performance in downstream tasks. According to findings from previous work [2], professionalism is a key indicator of high-quality opinions. Following this line of thought, we explore whether pre-finetuning with professionalism-aware tasks can enhance model performance in identifying high-quality and profitable investment suggestions [7].

The pre-finetuning task is designed to classify the source of a given analysis (a sentence), distinguishing between professionals and social media users, or to determine who (professionals or social media users) uses a given word more frequently.

6.2 Opinion Ranking

Table 6.3 Outline of the discussions on opinion components

	Our previous book [3]	This book
Target entity	Sect. 5.2	–
Market sentiment	Sect. 4.1	Sect. 2.2
Opinion holder	–	–
Publishing time	–	Sect. 2.3
Validity period of an opinion	–	Sects. 2.2 and 6.1
Market information set	–	Sect. 2.3
Analysis aspect	Sect. 4.1	–
Degree of market sentiment	–	Sect. 2.3
A set of claims	Sect. 4.1	Chap. 2
A set of premises	Sect. 4.1	Chap. 2
Opinion quality	Sect. 4.2	Sects. 2.3 and 6.2
Influence power	–	–

Table 6.4 MPP of Top 10% opinions with different methods [4, 7]

	Professional report (%)	Social media (%)
PLM (Best among BERT, Mengzi-Fin, and SCQF)	14.14	19.26
Forward-looking argument mining	**15.59**	19.41
Professionalism-aware pre-finetuning	14.62	**23.39**

We conduct experiments with three PLMs: BERT, Mengzi-Fin [18], and SCQF.[2] The MPP is used for evaluation, and the results are presented in Table 6.4.

The results indicate that professionalism-aware pre-finetuning is effective in ranking both professional reports and social media posts, outperforming the best-performing PLM. We also compare the performances with the forward-looking argument mining approach proposed in Chap. 2 on the same task. The findings suggest that, for ranking professional reports, the proposed argument mining approach performs better than pre-finetuning. However, when ranking social media posts, professionalism emerges as a highly significant feature.

In conclusion, while sentiment analysis and opinion mining have been widely researched, recent advancements have provided more fine-grained discussions, as shown in Table 6.3. Nonetheless, several tasks remain open for future research. We hope our discussions will encourage future work to explore avenues beyond sentiment analysis in financial NLP and expand discussions on topics other than market information prediction.

---

[2] https://huggingface.co/DMetaSoul/sbert-chinese-qmc-finance-v1.

6.3 Numeracy and Reasoning

As LLMs continue to demonstrate remarkable versatility across various domains, a key area of research is understanding their numerical reasoning capabilities and how these skills can be systematically enhanced. Numerical reasoning, encompassing tasks such as quantitative prediction, comparison, and inference, is foundational for a broad range of applications in science, finance, and everyday decision-making. Despite the advancements in pre-training and fine-tuning methods, effectively equipping models with robust numeracy remains a challenging and underexplored area.

Recent efforts in NLP research have introduced benchmarks and datasets that explicitly evaluate numeracy and reasoning skills. These include Numeracy-600K [5], which assesses models on number prediction tasks; EQUATE [14], which focuses on arithmetic and equation solving; and NumGLUE [12], a suite of tasks for testing quantitative understanding in broader linguistic contexts. Alongside these datasets, various methodological approaches, such as task-specific pre-finetuning and multi-scale modeling, have emerged as promising strategies for addressing these challenges [6].

Beyond simple numeracy, reasoning tasks often involve higher-order skills like multi-hop reasoning, data querying, and synthesis across disparate information sources. While traditional reasoning benchmarks, such as QA datasets for finance [9, 10, 19], offer valuable insights into domain-specific reasoning, they frequently assume structured input or only provide direct access to relevant data. This highlights a critical gap: the need for datasets and task formulations that mirror real-world workflows, where reasoning involves complex interactions between data retrieval and multi-step processing.

In this section, we focus on numeracy and reasoning, presenting our exploration of how pre-finetuning enhances these abilities. First, we analyze the impact of number comparison tasks on improving model numeracy and generalization across quantitative tasks. Subsequently, we introduce new findings from reasoning tasks that combine data querying and logic-driven processes, supported by our pilot dataset for database querying and reasoning (DBQR) [13]. By benchmarking state-of-the-art models, we identify current limitations in multi-hop reasoning and propose future research directions, including the refinement of multi-agent frameworks and the development of adaptive, task-specific model scales.

Firstly, continuing the discussion on pre-finetuning, we provide evidence supporting the usefulness of a pre-finetuning scheme for financial tasks in Sects. 3.1 and 6.2. In addition to its benefits for financial tasks, we observe that pre-finetuning can improve the general capabilities of models. Consequently, in this section, we focus on the numeracy of models, specifically addressing quantitative tasks such as Quantitative Prediction (QP), Quantitative Natural Language Inference (QNLI), and Quantitative Question Answering (QQA). For our experiments, we utilize Numeracy-600K [5], EQUATE [14], and NumGLUE Task 3 [12].

We propose using number comparison as a pre-finetuning task to enhance model numeracy [6]. Taking RoBERTa as an example, we present the performance changes

6.3 Numeracy and Reasoning

Table 6.5 Performance difference after using the pre-finetuning scheme

	QP		QNLI					QQA (%)
	Comment	Headline (%)	RTE-QUANT (%)	AWP-NLI (%)	NEWSNLI (%)	REDDITNLI (%)	Stress test (%)	
RoBERTa	26.40	19.26	2.37	−0.94	−0.76	5.52	1.01	−1.25

in Table 6.5. The performance gain in quantitative prediction is more substantial than in other tasks, with improvement observed in three out of five QNLI subsets. These results support the assertion that pre-finetuning can lead to significant performance gains, although slight detriments may occur in some cases.

Beyond simple numeracy tasks, LLMs are capable of tackling more complex reasoning tasks. As discussed in Chap. 4, simulating human workflows is a key direction for utilizing AI agents. Although there are several tabular QA datasets available in the financial domain [9, 10, 19], no dataset has been proposed for solving reasoning tasks from scratch, which involves querying data followed by reasoning. To address this gap, we introduce the first pilot dataset for this task setting [13]. Our experimental results with state-of-the-art LLMs indicate that while these models perform well on simple reasoning questions (e.g., extracting a few numbers for straightforward calculations) and even in conversation mode (requiring co-reference abilities), they struggle significantly with more complex tasks. Specifically, when multi-hop reasoning over multiple tables is required, models tend to fail in either the querying or reasoning steps, with querying being particularly problematic. Future studies could improve performance by enhancing the multi-agent interaction framework or employing multi-scale models to divide tasks into manageable subtasks, each addressed by models of different scales.

In summary, this section reviews our discussion on pre-finetuning. The scheme not only benefits financial tasks but also enhances the general capabilities of models. It shows the value of human insights and highlights its potential in feature engineering for model training. However, while LLMs perform well on traditional numerical reasoning tasks, their performance is weaker on real-world querying and reasoning tasks. This suggests that, although many efforts have been made to create benchmarks from existing datasets, a more pressing issue may be the development of new tasks that closely resemble real-world human workflows. To conclude, this section and our previous discussions show the significance of novel human insights and the simulation of human workflows. Future research may explore methods for the automatic discovery of meaningful features and the automatic planning of agent workflows.

6.4 Creative Agent

The advancement of AI agents has unlocked a wide range of applications across domains, demonstrating their capabilities in automating complex tasks, enhancing decision-making, and generating insights. AI agents are often lauded for their ability to synthesize vast amounts of data, identify patterns, and suggest novel solutions. However, studying how they help boost creativity hasn't been explored much in academic discussions. This section seeks to investigate the potential of AI agents in generating creative and practical contributions within research workflows.

The study of creativity in AI systems has typically focused on domains such as art, literature, and music, where subjective interpretation plays a significant role. Yet, in structured fields like finance and economics, creativity manifests through the ability to identify innovative methodologies, propose new research directions, and develop tools for complex problem-solving. Recent work suggests that AI agents can outperform human researchers in proposing unconventional ideas, though these ideas may often lack feasibility [16]. This duality–the generation of creative yet potentially impractical suggestions–underscores the need for systematic evaluation of the utility of AI-driven creativity in professional contexts.

One area ripe for exploration is the ability of LLMs to propose and refine keyword-based methodologies, which are widely employed in finance and economics. For example, keyword-based methods have been instrumental in constructing indices like the Economic Policy Uncertainty (EPU) index. Traditionally, human experts have had to manually identify these keywords through extensive reviews of news articles [1]. Given that LLMs are pre-trained on extensive corpora encompassing diverse sources, an intriguing question arises: can AI agents identify more effective or complementary keyword sets compared to those proposed by human experts?

This section provides a comparative analysis of human- and AI-suggested keyword sets and their respective effects on constructing economic indicators. Specifically, we examine the utility of LLM-generated keywords in developing EPU indices and evaluate their ability to explain and predict key economic variables [17]. Through this investigation, we aim to shed light on the broader implications of integrating AI agents into research workflows, particularly in domains where innovation and creativity play a pivotal role.

Our results show that the keywords suggested by LLMs differ significantly from those suggested by humans across data from five countries. LLMs suggest a larger number of terms than humans, while human keywords tend to include more country-specific terms, such as "The White House." Despite providing LLMs with definitions of the EPU index or assigning them various roles (e.g., editor, economist, minister of economics), the suggested keywords remain dissimilar to those proposed by humans. However, the aim of this discussion is not to replicate human-proposed keywords but to assess the usefulness of the LLM-suggested keywords. Consequently, we construct the EPU index based on the keywords generated by LLMs to evaluate its ability to explain or predict economic variables such as the non-farm employment rate, Consumer Price Index (CPI), Industrial Production Index (IPI), and average stock market price.

Table 6.6 Explainability and predictability of EPU indices with different keyword sets

	Explainability		Predictability	
	Human	LLM (best)	Human	LLM (best)
Non-farm	*	**	***	***
CPI				*
IPI	**		**	*
Stock market	***	*		

*, **, and *** denote significance under 95, 99, and 99.7% confidence levels, respectively [17]

Table 6.6 presents our findings. While human-suggested keywords demonstrate superior overall performance, LLM-suggested keywords exhibit considerable explainability and predictability for some economic variables. Beyond identifying this gap, we aim to highlight the potential of LLMs in economic and financial research. In our experiment, we shared only the concept of the EPU index with the LLM and obtained results rapidly. Compared to the manual process of data review and keyword suggestion by humans, LLMs offer an alternative approach for textual-based social science research to validate ideas. Although evaluating creativity remains difficult, traditional methods in social science can assess the usefulness of LLM outputs without necessitating human annotation. For example, developing a global EPU index covering approximately 200 countries would pose a significant challenge for experts in tailoring keyword sets across different languages and countries. In this context, AI agents are a suitable option, as they can suggest reasonable keywords for such research. Social scientists can then focus on analyzing the constructed index.

While our initial explorations don't assess creativity, this section encourages future research and workflows to focus on human-agent collaboration. Tasks could extend beyond automation (e.g., classification or summarization), with AI agents offering deeper insights into daily human tasks, even within professional contexts. In summary, we emphasize the importance of considering how agent-based frameworks can transcend knowledge limitations, with creativity as one of the most critical topics.

6.5 Summary

This chapter has explored various application scenarios of generative AI in the financial field, building upon concepts and methodologies discussed in previous sections. Focusing on the roles of impact duration, opinion ranking, numeracy and reasoning, and the creativity of AI agents, we provided comprehensive discussions on how generative AI is being harnessed in financial contexts. We emphasized the importance of incorporating human insights into AI workflows, leveraging retrieval-based approaches, and considering creativity in AI applications. We also highlighted the potential for future research to further expand these discussions, exploring the synergy between AI and human expertise to enhance financial decision-making and creativity in real-world applications.

References

1. BAKER, S. R., BLOOM, N., AND DAVIS, S. J. Measuring Economic Policy Uncertainty. *The Quarterly Journal of Economics 131*, 4 (2016), 1593–1636.
2. CHEN, C.-C., HUANG, H.-H., AND CHEN, H.-H. Evaluating the rationales of amateur investors. In *The World Wide Web Conference* (2021).
3. CHEN, C.-C., HUANG, H.-H., AND CHEN, H.-H. *From opinion mining to financial argument mining*. Springer Nature, 2021.
4. CHEN, C.-C., HUANG, H.-H., CHEN, H.-H., TAKAMURA, H., KOBAYASHI, I., AND MIYAO, Y. Enhancing investment opinion ranking through argument-based sentiment analysis. In *arXiv* (2024).
5. CHEN, C.-C., HUANG, H.-H., TAKAMURA, H., AND CHEN, H.-H. Numeracy-600K: Learning numeracy for detecting exaggerated information in market comments. In *Proceedings of the Fifty-Seventh Annual Meeting of the Association for Computational Linguistics* (Florence, Italy, July 2019), Association for Computational Linguistics, pp. 6307–6313.
6. CHEN, C.-C., TAKAMURA, H., KOBAYASHI, I., AND MIYAO, Y. Improving numeracy by input reframing and quantitative pre-finetuning task. In *Findings of the Association for Computational Linguistics: EACL 2023* (Dubrovnik, Croatia, May 2023), A. Vlachos and I. Augenstein, Eds., Association for Computational Linguistics, pp. 69–77.
7. CHEN, C.-C., TAKAMURA, H., KOBAYASHI, I., AND MIYAO, Y. Professionalism-aware pre-finetuning for profitability ranking. In *Proceedings of the 33rd ACM International Conference on Information & Knowledge Management* (2024).
8. CHEN, C.-C., TSENG, Y.-M., KANG, J., LHUISSIER, A., SEKI, Y., LEE, H., DAY, M.-Y., TU, T.-T., AND CHEN, H.-H. Multi-lingual ESG impact duration inference. In *Proceedings of the Joint Workshop of the 7th Financial Technology and Natural Language Processing, the 5th Knowledge Discovery from Unstructured Data in Financial Services, and the 4th Workshop on Economics and Natural Language Processing* (Torino, Italia, May 2024), C.-C. Chen, X. Liu, U. Hahn, A. Nourbakhsh, Z. Ma, C. Smiley, V. Hoste, S. R. Das, M. Li, M. Ghassemi, H.-H. Huang, H. Takamura, and H.-H. Chen, Eds., Association for Computational Linguistics, pp. 219–227.
9. CHEN, Z., CHEN, W., SMILEY, C., SHAH, S., BOROVA, I., LANGDON, D., MOUSSA, R., BEANE, M., HUANG, T.-H., ROUTLEDGE, B., AND WANG, W. Y. FinQA: A dataset of numerical reasoning over financial data. In *Proceedings of the 2021 Conference on Empirical Methods in Natural Language Processing* (Online and Punta Cana, Dominican Republic, Nov. 2021), Association for Computational Linguistics, pp. 3697–3711.
10. CHEN, Z., LI, S., SMILEY, C., MA, Z., SHAH, S., AND WANG, W. Y. ConvFinQA: Exploring the chain of numerical reasoning in conversational finance question answering. In *Proceedings of the 2022 Conference on Empirical Methods in Natural Language Processing* (Abu Dhabi, United Arab Emirates, Dec. 2022), Y. Goldberg, Z. Kozareva, and Y. Zhang, Eds., Association for Computational Linguistics, pp. 6279–6292.
11. DONG, Q., LI, L., DAI, D., ZHENG, C., WU, Z., CHANG, B., SUN, X., XU, J., AND SUI, Z. A survey on in-context learning. *arXiv preprint* arXiv:2301.00234 (2022).
12. MISHRA, S., MITRA, A., VARSHNEY, N., SACHDEVA, B., CLARK, P., BARAL, C., AND KALYAN, A. NumGLUE: A suite of fundamental yet challenging mathematical reasoning

References

tasks. In *Proceedings of the 60th Annual Meeting of the Association for Computational Linguistics (Volume 1: Long Papers)* (Dublin, Ireland, May 2022), Association for Computational Linguistics, pp. 3505–3523.

13. NARARATWONG, R., CHEN, C.-C., KERTKEIDKACHORN, N., TAKAMURA, H., AND ICHISE, R. DBQR-QA: A question answering dataset on a hybrid of database querying and reasoning. In *Findings of the Association for Computational Linguistics ACL 2024* (Bangkok, Thailand and virtual meeting, Aug. 2024), L.-W. Ku, A. Martins, and V. Srikumar, Eds., Association for Computational Linguistics, pp. 15169–15182.
14. RAVICHANDER, A., NAIK, A., ROSE, C., AND HOVY, E. EQUATE: A benchmark evaluation framework for quantitative reasoning in natural language inference. In *Proceedings of the 23rd Conference on Computational Natural Language Learning (CoNLL)* (Hong Kong, China, Nov. 2019), Association for Computational Linguistics, pp. 349–361.
15. SEKI, Y., SHU, H., LHUISSIER, A., LEE, H., KANG, J., DAY, M.-Y., AND CHEN, C.-C. Ml-promise: A multilingual dataset for corporate promise verification. In *arXiv* (2024).
16. SI, C., YANG, D., AND HASHIMOTO, T. Can llms generate novel research ideas? a large-scale human study with 100+ nlp researchers. *arXiv preprint* arXiv:2409.04109 (2024).
17. YEH, H.-H., HUANG, Y.-L., PARK, Z., AND CHEN, C.-C. Automation of text-based economic indicator construction: A pilot exploration on economic policy uncertainty index. In *CIKM* (2024), vol. 24, pp. 21–25.
18. ZHANG, Z., ZHANG, H., CHEN, K., GUO, Y., HUA, J., WANG, Y., AND ZHOU, M. Mengzi: Towards lightweight yet ingenious pre-trained models for chinese. *arXiv preprint* arXiv:2110.06696 (2021).
19. ZHU, F., LEI, W., HUANG, Y., WANG, C., ZHANG, S., LV, J., FENG, F., AND CHUA, T.-S. TAT-QA: A question answering benchmark on a hybrid of tabular and textual content in finance. In *Proceedings of the 59th Annual Meeting of the Association for Computational Linguistics and the 11th International Joint Conference on Natural Language Processing (Volume 1: Long Papers)* (Online, Aug. 2021), Association for Computational Linguistics, pp. 3277–3287.

Open Access This chapter is licensed under the terms of the Creative Commons Attribution 4.0 International License (http://creativecommons.org/licenses/by/4.0/), which permits use, sharing, adaptation, distribution and reproduction in any medium or format, as long as you give appropriate credit to the original author(s) and the source, provide a link to the Creative Commons license and indicate if changes were made.

The images or other third party material in this chapter are included in the chapter's Creative Commons license, unless indicated otherwise in a credit line to the material. If material is not included in the chapter's Creative Commons license and your intended use is not permitted by statutory regulation or exceeds the permitted use, you will need to obtain permission directly from the copyright holder.

Chapter 7
Looking to the Future

In this book, we have first continued the discussion from our previous work, focusing on financial argument mining concepts, specifically forward-looking argument mining. We then explored recent trends in agent-based approaches, including single-agent design, multi-agent interactions, and multi-scale model synergy. Several novel application scenarios were examined, such as multiple question generation, LLM agent-based modeling, and opinion ranking. In this final chapter, we revisit the connection between this book and the future directions proposed in 2021, analyzing what has been addressed (explored) and what remains unresolved. We also discuss our investigation into how LLMs influence human (expert) decisions. Several future research directions are proposed in Sect. 7.1, and the book is concluded in Sect. 7.3.

7.1 Progress on Previously Proposed Research Directions

The development of financial argument mining and related research areas has seen significant progress since the publication of our previous book. In this section, we revisit the directions proposed earlier and assess advancements made in this book, while identifying unresolved challenges that require further investigation. Table 7.1 systematically links previously suggested topics to their corresponding discussions in this book. Our goal is to provide a cohesive narrative on the state of progress and outline promising paths forward.

We first address opinion mining and argument mining (R1–R6), where notable progress has been made in exploring the validity period of financial opinions, the quality of argumentation, and the structural relations between argumentative units. Chapter 2 elaborates on these advancements, and Sect. 6.2 includes pairwise opinion ranking. However, analyzing inter-opinion relationships remains an open issue. To bridge this gap, we here introduce novel datasets and methodologies for studying the distribution of relationships within social media posts, along with an analysis of how LLM-generated financial reports influence human decision-making.

Table 7.1 Link with the research directions suggested in our previous book [2]

From opinion mining to financial argument mining [2]			This book
Index	Section	Research topic	Section
R1	2.1	Extracting/estimating the validity period of a financial opinion	2.2 and 6.1
R2	2.2	Relation linking for elementary argumentative units in a financial opinion	2.1
R3	2.3	Analyzing relations between financial opinions	7.1
R4	4.2	Evaluating the quality of a financial opinion	2.3 and 6.2
R5	4.3	Estimating the influence of a financial opinion	–
R6	4.3	Implicit information inference	–
R7	5.2	General numeral attachment in financial narratives	3.1
R8	5.3	Exploring model numeracy	6.3
R9	6.1	Detection of false financial information	–
R10	6.1	Generation of financial analysis reports	7.1
R11	6.2	Financial opinion-based personalized recommendation	–
R12	6.3	Improving services for both employees and customers	3.2 and 5.2
R13	7.1	Organizing multimodal financial data	4.1 and 7.1
R14	7.1	Applying the proposed structures to other domains	–

In the domain of numeracy and reasoning (R7–R9), Sects. 3.1 and 6.3 provide fresh insights. While we have extended discussions on services for employees and customers (R12) by introducing new application scenarios, such as audience question generation and speech script refinement, certain topics, like personalized recommendations (R11), are not emphasized in this book. Instead, these areas remain ripe for future exploration.

Furthermore, advancements in multimodal data processing (R13) are preliminarily discussed, showcasing how financial insight generation can benefit from integrating tabular data, price charts, and textual narratives. This exploration lays the groundwork for extending insights to other domains (R14), inviting researchers to evaluate the scalability and applicability of proposed frameworks in diverse NLP application scenarios.

Finally, while the influence of LLM-generated analyses on human decision-making presents various opportunities, it also raises concerns about compliance and misinformation (R9). As highlighted in Sects. 4.1 and 4.3, human-centered evaluation methods are essential for addressing these challenges. Future research should prioritize compliance-oriented frameworks and advanced multimodal modeling techniques to enhance financial insight generation, as summarized in Table 7.3.

7.1 Progress on Previously Proposed Research Directions

This section provides a comprehensive review of previously proposed research directions, discussing advancements, challenges, and pathways for future exploration. Our hope is that these insights will inspire ongoing efforts to advance the field of financial argument mining and NL-driven agent-based modeling applications.

Let us start with extending the relationship between financial opinions. We propose a dataset to capture the relationship between two social media posts on a financial social media platform [3, 8]. When given two posts, we provide "support," "attack," or "neutral" labels to describe their relationship. The distribution of these labels is 56.47, 33.10, and 10.43%, respectively. This indicates that social media users frequently respond to posts they agree with, while only 33.10% of opinion pairs are labeled as "attack". Our experimental results show that we can improve the performance by 7.3% in macro-F1 scores through multilingual cross-source training, which can be considered a form of data augmentation. Lin et al. [7] proposed a two-step transformer architecture to address this task, achieving state-of-the-art performance by improving the macro-F1 score by approximately 1.6%. Future studies can adopt this task as a pre-finetuning stage for downstream financial tasks, such as opinion ranking or market information prediction.

In addition to analyzing opinions between humans, we explore the relationship between opinions expressed by agents and humans. Specifically, we examine whether an agent's opinion influences human decision-making [9]. While LLMs can generate fluent content, their persuasiveness and potential to shape human thought remain open research questions. In Sect. 4.3, our simulation assumes that all individuals follow the AI agent's suggestions. Here, we discuss the extent to which humans are likely to follow AI agents' suggestions in financial decision-making scenarios.

Our experimental setup is as follows:

- We use GPT-4 to generate (1) a summary, (2) an analysis (given stance), and (3) a promotional analysis (given stance) based on the transcript of an earnings call.
- We invite participants from three categories: amateurs, experts (working in the financial industry), and veteran experts (with over 10 years of experience in the financial industry).
- The decision-making process consists of two rounds. In the first round, participants make a three-day trading decision based on the provided summary. In the second round, they receive a (promotional) analysis with stance and decide whether to modify their initial decision.
- Participants receive an hourly salary that is 1.5 times their original rate if they make correct decisions for over 50% of instances.

Table 7.2 presents the experimental results. Firstly, GPT-4's analysis influences only a small number of human decisions, with the smallest impact observed among veteran experts. The frequency of decision changes among amateurs is double that of veteran experts. Our in-depth analysis reveals that promotional analysis is perceived as more convincing, logical, and useful by all participants. This finding highlights the

Table 7.2 GPT-4 influence on human decisions

	Amateur (%)	Expert (%)	Veteran (%)
Frequency	31.30	24.70	15.60
Decrease of accuracy	15.40	16.60	11.10

first risk associated with using large language models (LLMs) to generate financial analysis or investment suggestions. In the financial market, promoting investment products requires caution due to the strict regulations governing financial product promotion across different regions. In many jurisdictions, compliance with specific laws and regulations established by regulatory authorities is necessary. For example, regulatory authorities generally mandate that when promoting investment products, information must be complete, accurate, and non-misleading. This includes comprehensive disclosure of investment risks, expected returns, product terms, and other details to ensure that investors can make informed decisions. Furthermore, advertisements for financial products must not exaggerate facts, guarantee returns, or contain unsubstantiated claims; they must adhere to principles of objectivity, balance, and fairness. Most importantly, fraudulent activities are strictly prohibited across all financial markets. During the promotion of investment products, any form of fraud, misleading conduct, manipulation, or concealment of truth is forbidden. This is closely related to the research topic of false financial information (R9) as outlined in Table 7.1. In summary, it is possible to legally promote investment products within the financial market, but it is crucial to comply with regulatory rules to protect investors and maintain fairness and transparency. We urge the research community to focus on generating insights for financial investments to pay careful attention to compliance and false information research topics in the financial domain.

An additional noteworthy issue in Table 7.2 is that GPT-4-generated analysis negatively affects the accuracy of both amateurs and experts. These findings indicate that although GPT-4 can produce persuasive analysis, it may not necessarily support humans in making decisions. This raises a research issue regarding evaluation. In Sect. 4.1, we evaluated the generated analysis based on its usefulness and insightfulness. Here, we assess whether the generated analysis alters individual decisions and whether it leads to improved decision-making. Unlike the design of market prediction models (buy/sell), the aim of generating analysis should be to provide individuals with comprehensive information for making informed final decisions. For example, while discussions on analysis generation exist, the generation of risk reminders, as mentioned in Sect. 5.1, remains underexplored. We encourage a focus on generating reports aimed at human-centered decision-making. Beyond traditional evaluation metrics and human assessments, the evaluation of forward-looking statements should incorporate a Human-Computer Interaction (HCI) style of assessment, such as the think-aloud protocol [6] and cognitive walkthrough [10]. In conclusion, our experiments in Sect. 5.1 and in this section demonstrate that, while agents are capable of generating reasonable analysis, it is not clear whether this analysis

7.2 Future Directions

can effectively assist human decision-making. We encourage future research in this domain to consider our recommendations on evaluation methods for human-centered decision-making.

In this book, our discussions on insight generation center on earnings calls, where the input data may be either text or audio. However, much of the data in the financial field is presented in multimodal formats, such as tabular data [11], images [1], and price charts [5]. We provide preliminary explorations on generating claims based on given premises [4]. Our findings indicate that it remains challenging for models to learn from past presentations to select the correct topic for claim generation as company managers. Additionally, we explore insight-generation tasks based on either a given news article or a price chart [5]. While PLMs can learn narrative styles from training data, they often produce fine-grained errors involving numbers and causal relationships between financial terms. Future studies can investigate these tasks using a combination of a multi-agent framework and a multi-scale model approach to achieve improved performance. For example, rather than employing a single agent or model to generate entire content directly, regression models could be used for numerical predictions, which would then inform subsequent content generation. This approach parallels the workflow of professional analysts: instead of drafting a report immediately after gathering information, some analysts first infer numerical values—such as EPS, sales, and costs—based on the available information, and then compose the report based on their beliefs, analyses, and inferences.

In summary, this section has reviewed the progress on previously proposed research directions. We highlighted which topics have been addressed in the current book and which remain open. Advancements in opinion mining and argument mining were discussed, including explorations of validity periods, argument units, and argument quality. While some issues, like the analysis of relationships among financial opinions, remain open, we introduced a new dataset for capturing these relationships and examined how LLMs like GPT-4 influence human decision-making. Findings indicate that LLM-generated analyses can affect decisions but may not improve accuracy, implying potential risks related to compliance and misinformation. We have thus emphasized the need for human-centered evaluation methods and suggested future research directions involving multimodal data and advanced modeling techniques to enhance financial insight generation.

7.2 Future Directions

Table 7.3 summarizes the future research directions that we suggest and plan to explore in the near future. By comparing these directions with those in Table 7.1, it is apparent that the focus has shifted towards more subjective topics that extend beyond single model design or feature extraction. On one hand, this book emphasizes forward-looking arguments; on the other, with advancements in AI agents, we are now able to address more complex tasks. In this section, we review these research directions individually and share our thoughts.

Table 7.3 Summary of future research directions in Agent AI for finance

Index	Research topic	Section
D1	Scenario verification	2.2
D2	Forecasting skill assessment	2.3
D3	New feature discover	3.1
D4	Presentation preparation	3.2
D5	Organization simulation	4.2
D6	Society simulation	4.3
D7	Expressing timing identification	5.1
D8	Design of dynamic interaction loop	5.2
D9	Application of impact duration	6.1
D10	Opinion ranking	6.2
D11	Agent AI for creative idea	6.4
D12	Generated insight evaluation	7.1
D13	Compliance checking	7.1
D14	Agent AI design	7.2

In Chap. 2, we discussed the rationales underlying financial argument mining and proposed an additional argument unit, the scenario, to distinguish forward-looking inferences from events that have already occurred (premises). We further explored the application of argument mining to assess the quality and forecasting ability of financial opinions. Here, we wish to highlight two research directions based on prior discussions. The first is scenario verification (D1). In our previous book, we proposed comparing figures within an opinion to market data to capture fine-grained sentiment degrees. We also suggested verifying whether such fine-grained predictions hold true in the future. For example, if investors provide price targets indicating their expectations of price movement, it is possible to verify these expectations retrospectively. This type of verification constitutes claim verification. However, verifying a scenario is more complex. For example, analysts might present the following analysis:

> Based on a certain premise, *event A* may occur and subsequently influence the operation of the company from *aspect B*. This, in turn, may ultimately affect *account C*. Thus, we anticipate the 25Q1 EPS may decrease by *D%*.

Here, D can be verified straightforwardly by extracting the number (D), the time (25Q1), and the associated entity (EPS), and then comparing them with the correct figures in the future. However, the scenario composed of *event A*, *aspect B*, and *account C* is significantly more complex. For example, it may be challenging to verify whether *event A* truly affects the company's operations from *aspect B*, and whether *account C* is genuinely influenced by *event A*. We term this task "scenario verification." Although we proposed a dataset in Sect. 6.1 to identify when managers' ESG promises should be verified, the topic of scenario verification remains

7.2 Future Directions

largely unexplored. Consequently, we suggest that future research should focus on developing more fine-grained approaches to scenario and promise verification.

The second research direction we highlight from the discussions in Chap. 2 is forecasting skill assessment. Although some progress has been made in previous studies, the definitions of quality and forecasting skills remain diverse. Moreover, whether high-quality analysis necessarily correlates with high forecasting skills is still an open question. Although such assessments are inherently subjective, they are crucial, particularly when discussions focus on future outcomes. For example, evaluation metrics exist for professional analysts in financial institutions, and the extent to which these can be applied to the evaluation of agent-generated content presents an interesting research avenue. This is also related to the generated insight evaluation task (D12). We leave these topics for future research to explore.

In Sect. 3.1, we discussed agents and models that learn from human insights. The next step involves the automatic discovery of new features (D3) and creativity (D11). While achieving this may still be challenging for a single agent or model, we believe it is feasible within the Agent AI framework by combining multiple agents, models, and tools (such as retrieval mechanisms). For example, good ideas often emerge after several rounds of discussion and multiple stages of thought. Based on this concept, multiple rounds of discussion and multi-scale modeling could be an effective approach to refine the agent's output. We can establish various objectives (e.g., forecasting skill, reasonability), but creativity is, in our view, the most critical, as it can lead to new ideas and provide alternative insights from historical experiences. The development of creative Agent AI will elevate human intelligence to a new level. However, creativity is a subjective quality, and automating its evaluation is one of the subtasks we need to address in this research direction. As discussions are still in the early stages, we aim to emphasize this as a direction for future work in the development of Agent AI.

The next topic is human behavior simulation in Chap. 4. It addresses the behavior of an individual (D4), the interaction of an organization (D5), and the simulation of a society (D6). For the individual, we propose focusing on text generation for presentation preparation (D4). Although it remains an NLP task, it encompasses a wide range of topics. For example, audience question generation and rehearsing answers can be applied in educational and professional presentation preparation scenarios. Moreover, refining the presentation script based on anticipated audience feedback aligns with the workflow of a human's working process. The document refinement discussed in Sect. 5.2 also falls within this scope. Consequently, we suggest that presentation preparation tasks could become standard tasks for future NLP frameworks, though their evaluation remains a challenge. A more fine-grained approach could be introduced when addressing this task. For example, managers may attempt to mitigate large fluctuations in their company's stock prices while preparing presentations.

Taking this further, organization simulation (D5) involves a more complex discussion of multi-agent interaction. This could entail simulating a professional organization, but it could also lead to the creation of a new organizational structure within the multi-agent framework. The ultimate goal is to simulate the behavior and decision-making processes of an entire society (D6), a key objective of agent-based modeling.

When a new policy is introduced, we seek to simulate its potential impacts—who will be affected, and to what degree will the policy meet its original objectives? These are critical questions we need to address through simulation. Moving beyond scenario planning and generating textual explanations, the use of agent interaction to simulate societal behavior for addressing such questions presents a significant and challenging task.

Returning to the example of the financial market, after an earnings call, rather than drafting reports, could we employ a multi-agent framework to simulate the interactions of market participants the following day? There are numerous intriguing tasks and topics to explore in this area, and we wish to emphasize this research direction for future studies.

In the financial field, temporality is a crucial subject. Timing identification (D7) and the application of impact duration (D9) are directly related to this topic. In simulations of an organization or society, agents need to select an appropriate time to express their opinions or take actions. To achieve better simulation results, addressing timing identification is essential. As mentioned in several sections, impact duration is a new concept in opinion mining. Previous studies have aggregated opinions (sentiments) over various periods to predict market information. However, would conducting fine-grained analyses and understanding the sentiment across different time periods lead to improved outcomes? The analysis and application of opinion quality evaluation and ranking (D10) follow a similar logic. Should we consider all opinions, or should we prioritize high-quality ones? Should the same evaluation metrics be applied to analyses over different durations? Should the focus vary? While we have explored a few tasks from these perspectives, there remains room for discussion on how these fine-grained opinion components will influence downstream tasks. It is worth noting that the opinion components discussed in our previous book remain relevant in the era of LLMs. However, the focus has shifted from extracting these components to applying them effectively.

The Dynamic Interaction Loop (D8) presents a novel approach distinct from the currently popular multi-agent framework, emphasizing the necessity of multi-scale models to achieve synergy. We consider this framework an extension of the multi-agent paradigm, which could lead to improved outcomes. However, there have been limited explorations based on this concept. We hope future research will build on our discussions in Sect. 5.2 to expand and refine this framework, and further enable it to tackle more complex tasks.

In terms of opinion ranking (D10), we use return as a proxy in this book. However, numerous other concepts could be employed to rank forward-looking opinions. For example, professionalism, as previously discussed, serves as a basis for ranking opinions. Whether such factors influence human decision-making and lead to better outcomes warrants further exploration. This also ties into the remaining topic of influence power (R5), as discussed in Table 7.1. Future research can also address evaluation issues in relation to the estimation of influence power. The stagnation of research in this direction is partly due to the lack of exploration in the automatic design of evaluation metrics and insufficient discussion on the subject. We reiterate the importance of performing evaluations within the context of human-computer

7.2 Future Directions

interaction. In addition to traditional metrics, the effectiveness of the generated content in aiding human decision-making is crucial.

Building on the theme of evaluation, the evaluation of generated insight (content) (D12) is closely related to (D10). In this context, we would like to introduce the concept of "surprise estimation." In an efficient market, all available information is reflected, and any surprises would result in unexpected movements. When a professional's presentation or analysis diverges from AI-generated content, it may result in surprises. For example, if agents are provided with all available information and trained on all historical data, their outputs could be regarded as optimized statistical outcomes. Suppose all individuals rely on Agent AI for decisions regarding the upcoming FED presentation; the agent's output could represent the collective view on the matter. If the FED aims to introduce a surprise to the market, they might adjust their presentation accordingly. On the other hand, if the FED's presentation significantly differs from the agent-generated content, this could suggest transparency issues, implying that some information utilized in decision-making has not been disclosed to the public. While this topic remains underexplored, we highlight it here to encourage a reevaluation of the implications of agent-generated content.

The compliance checking (D13) issue, as discussed in Sect. 7.1, is a significant concern. Although numerous approaches exist to prevent agents from generating noncompliant content, there are also techniques for jailbreaking these restrictions. Furthermore, compliance is especially crucial in the financial industry due to its high level of regulation. Every output (document) must be carefully examined and aligned with specific guidelines. From an influence perspective, we have already demonstrated that generated content poses a potential risk of misuse and can impact human decision-making. In addition to quality assessment, we consider compliance checking to be an important task for mitigating such risks. Moreover, automatic compliance checking can enhance employee efficiency by highlighting potential oversights in the manual review process.

Returning to the topic of Agent AI (D14), as illustrated in Fig. 1.1, the Agent AI framework is envisioned to include a leading agent responsible for agent or model selection, thereby enabling a multi-scale model synergy approach. This leading agent would also oversee the interactions between agents and models and serve as the primary interface for communication with users. In this setup, all agents and models function as team members under the leadership of this main agent, with tasks being distributed and assigned to the appropriate agent or model. We believe this all-in-one design represents the future of personal AI assistants, and we hope the discussions in this book inspire readers to design Agent AI systems for the financial sector.

To develop the Agent AI system, commentary generation represents a critical component. This capability allows AI agents to produce contextually relevant, insightful, and actionable feedback or explanations during their operation. For example, the multi-agent frameworks discussed in Chap. 4 currently rely solely on the basic functionality of LLMs without specific tuning for professional commentary. If the annotator agents (Sect. 4.1) could generate commentary or engage in debates with one another, the results could potentially improve. Similarly, if the head trader (Sect. 4.2) could provide commentary explaining its rejection of certain trading decisions, it would contribute to more refined trading strategies through iterative discussions. Additionally, enabling the LLM-agent (Sect. 4.3) to engage in discussions with neighbors and base decisions on these interactions could produce simulation outcomes more closely aligned with real-world scenarios.

Moreover, the leading agent within the Agent AI system should be equipped to generate commentary on each reply and action, further determining subsequent steps or terminating discussions as necessary. For example, the leading agent could explain its prioritization of issues based on urgency and user impact. This functionality is pivotal in enhancing the transparency and interpretability of the Agent AI's decision-making processes, thereby fostering user trust and engagement. Commentary generation also enables the leading agent to more effectively manage subordinate agents by providing clear rationales for task assignments, performance evaluations, and inter-agent communication strategies.

One of the primary reasons commentary generation is indispensable lies in its ability to bridge the gap between complex machine operations and human understanding. By producing detailed commentaries, the Agent AI can articulate the reasoning behind its decisions, outline alternative strategies, and summarize outcomes in a way that is accessible to human users. For example, in a financial context, it might analyze risk levels and expected returns to explain why a specific investment was recommended. This not only empowers users to make informed decisions but also facilitates error analysis and system debugging.

The applications of commentary generation are diverse. In real-time logistics management, the Agent AI could narrate its observations and decisions, such as rerouting deliveries to avoid traffic congestion or adjusting schedules in response to unforeseen delays. In collaborative tasks, commentary can enhance team coordination by providing regular progress updates and identifying potential bottlenecks. In training and educational contexts, it can elucidate complex workflows or concepts. For example, in a programming training environment, the system could explain an algorithm step-by-step or justify specific debugging approaches. Similarly, in the financial sector, it could offer detailed explanations for investment strategies, risk evaluations, or regulatory compliance measures.

In conclusion, commentary generation not only enhances the operational capabilities of the Agent AI but also strengthens the synergy between AI agents and users. By integrating this functionality into the Agent AI framework, the system becomes more powerful, accessible, and trustworthy.

7.3 Conclusion

Four years ago, our focus was on how to deeply understand financial texts, especially opinions, and pursue generating analyses and reports similar to those produced by analysts. In this book, we have already seen that AI agents can generate fluent analyses and even influence the decisions of experts. The rapid development of the NLP field leads us to think: what will the next step be?

7.3 Conclusion

In this book, we try to sort it out step-by-step. We first analyze the progress, extensions, and applications of the next step proposed four years ago, namely, financial argument mining. We then discuss the transformations and possibilities brought by AI agents, moving from the discussion of a single agent to multi-agent systems, presenting the limitless potential of generative AI. This ranges from simulating individual thinking to simulating organizational structures, and further extending to simulating societal operations. These scenarios bring fresh perspectives to financial NLP and move beyond merely extracting single pieces of information or generating individual documents. AI agents offer new hopes and challenges for simulation from the perspective of social sciences.

But does this mean we have already left the era of small models behind? Our view is that this is not the case. Instead, we should combine the models from both eras, leveraging their respective strengths to achieve synergy. This is what we refer to in this book as the multi-scale model dynamic interaction loop. We hope that the series of step-by-step discussions in this book can provide ideas and foundations for future research.

Many application scenarios, such as the importance of impact duration, the challenges of opinion ranking, more complex information retrieval and reasoning, and even the creative inspiration of agents, are still in the early stages of research. Achieving these goals and tasks through collaboration between agents and models is a pursuit we aim for in the near future.

In the final chapter, we utilized Table 7.1 to demonstrate that we have provided some preliminary answers and explorations regarding the "future directions" proposed four years ago. Additionally, we introduced a new set of future research directions in Table 7.3, which summarizes our considerations for the next steps. While this book primarily focuses on applications in finance, many of the future research directions are general in nature, and we believe they can be extended to other domains. For example, scenario planning and verification (D1) is widely applicable in geopolitics and business. Discovering new features (D3) and creative agent AI (D11) could also serve as topics in biomedical and other research fields. Presentation preparation (D4) is a general task, though we focus on financial presentations for our examples and experiments. Simulation methods (D5 and D6) can be applied to various domains, including economics and management. Timing identification (D7) is also crucial for agents across different fields; for example, a psychologist agent may need to carefully select the timing of their interventions. The concept of a dynamic interaction loop (D8) provides a framework that can be broadly applied. Compliance checking (D13) is likewise not restricted to the financial industry. Overall, we hope that our discussions, grounded in the field with which we are most familiar, will inspire future research that extends these ideas to other areas.

In conclusion, this book offers a comprehensive journey through the evolution of financial NLP, and reveals the immense potential of AI agents in reshaping the landscape of financial analysis and beyond. We have explored how AI can transcend the limitations of traditional models and seamlessly integrate both small and large-scale systems into a dynamic interaction loop that unlocks new possibilities. As we push the boundaries of what's possible—simulating thought processes, organizations, and even societal operations—the horizon of AI's role in finance and other domains continues to expand. Yet, this is only the beginning. The challenges we have outlined and the new research directions we have proposed represent the frontier of innovation. We stand on the edge of a new era where collaboration between agents and models holds the key to unlocking unprecedented insights and capabilities. We hope that the insights presented in this book will not only guide the future of financial NLP but also inspire advancements across other fields. We also hope that our thoughts will contribute a bit to driving forward the next wave of AI-driven discovery. The future awaits, and with it, limitless potential.

References

1. Babkin, P., Watson, W., Ma, Z., Cecchi, L., Raman, N., Nourbakhsh, A., and Shah, S. Bizgraphqa: A dataset for image-based inference over graph-structured diagrams from business domains. In *Proceedings of the 46th International ACM SIGIR Conference on Research and Development in Information Retrieval* (2023), pp. 2691–2700.
2. Chen, C.-C., Huang, H.-H., and Chen, H.-H. *From opinion mining to financial argument mining*. Springer Nature, 2021.
3. Chen, C.-C., Lin, C.-Y., Chiu, C.-J., Huang, H.-H., Alhamzeh, A., Huang, Y.-L., Takamura, H., and Chen, H.-H. Overview of the ntcir-17 finarg-1 task: Fine-grained argument understanding in financial analysis. In *Proceedings of the 17th NTCIR Conference on Evaluation of Information Access Technologies, Tokyo, Japan* (2023), pp. 12–15.
4. Chen, C.-C., and Takamura, H. Term-driven forward-looking claim synthesis in earnings calls. In *Proceedings of the 2024 Joint International Conference on Computational Linguistics, Language Resources and Evaluation (LREC-COLING 2024)* (Torino, Italia, May 2024), N. Calzolari, M.-Y. Kan, V. Hoste, A. Lenci, S. Sakti, and N. Xue, Eds., ELRA and ICCL, pp. 15752–15760.
5. Chen, C.-C., Takamura, H., Kobayashi, I., and Miyao, Y. Fingen: A dataset for argument generation in finance. *arXiv preprint* arXiv:2405.20708 (2024).
6. Ericsson, K. A., and Simon, H. A. Verbal reports as data. *Psychological review 87*, 3 (1980), 215.
7. Lin, Z. A., Li, H. M., Lin, A., Kao, Y. C., Hsu, C. S., and Fan, Y. C. Quack at the ntcir-17 finarg-1 task: Boosting and mlm enhanced financial knowledge sequence classification. In *Proceedings of the 17th NTCIR conference on evaluation of information access technologies*. https://doi.org/10.20736/0002001305 (2023).
8. Ruiz-Dolz, R., Chiu, C.-J., Chen, C.-C., Kando, N., and Chen, H.-H. Learning strategies for robust argument mining: An analysis of variations in language and domain. In *Proceedings of the 2024 Joint International Conference on Computational Linguistics, Language Resources and Evaluation (LREC-COLING 2024)* (Torino, Italia, May 2024), N. Calzolari, M.-Y. Kan, V. Hoste, A. Lenci, S. Sakti, and N. Xue, Eds., ELRA and ICCL, pp. 10286–10292.
9. Takayanagi, T., Takamura, H., Izumi, K., and Chen, C.-C. Beyond turing test: Can gpt-4 sway experts' decisions? *arXiv preprint* arXiv:2409.16710 (2024).

References

10. WHARTON, C., RIEMAN, J., LEWIS, C., AND POLSON, P. The cognitive walkthrough method: A practitioner's guide. In *Usability inspection methods*. 1994, pp. 105–140.
11. ZHU, F., LEI, W., HUANG, Y., WANG, C., ZHANG, S., LV, J., FENG, F., AND CHUA, T.- S. TAT-QA: A question answering benchmark on a hybrid of tabular and textual content in finance. In *Proceedings of the 59th Annual Meeting of the Association for Computational Linguistics and the 11th International Joint Conference on Natural Language Processing (Volume 1: Long Papers)* (Online, Aug. 2021), Association for Computational Linguistics, pp. 3277–3287.

Open Access This chapter is licensed under the terms of the Creative Commons Attribution 4.0 International License (http://creativecommons.org/licenses/by/4.0/), which permits use, sharing, adaptation, distribution and reproduction in any medium or format, as long as you give appropriate credit to the original author(s) and the source, provide a link to the Creative Commons license and indicate if changes were made.

The images or other third party material in this chapter are included in the chapter's Creative Commons license, unless indicated otherwise in a credit line to the material. If material is not included in the chapter's Creative Commons license and your intended use is not permitted by statutory regulation or exceeds the permitted use, you will need to obtain permission directly from the copyright holder.

The manufacturer's authorised representative in the EU is Springer Nature Customer Service Centre GmbH, Europaplatz 3, 69115 Heidelberg, Germany. If you have any concerns regarding our products, please contact ProductSafety@springernature.com

Printed and bound by CPI Group (UK) Ltd, Croydon, CR0 4YY
26/03/2026
02078992-0003